TVRs
Volume 1: Grantura to Taimar

TVRs

Volume 1: Grantura to Taimar

A collector's guide
by Graham Robson

MOTOR RACING PUBLICATIONS LTD
Unit 6, The Pilton Estate, 46 Pitlake, Croydon CR0 3RY, England

First published 1993

British Library Cataloguing in Publication Data

Robson, Graham
TVRs. – Vol.1: Grantura to Taimar. –
3Rev.ed. – (Collector's Guide Series)
I. Title II. Jackson, Anne III. Series
629.2222

ISBN 0-947981-80-2

Part of this book was previously published in 1987 by Motor Racing Publications in the Collector's Guide *The TVRs*, which this book and its companion volume (ISBN 0 947981 81 0) supersedes.

Typeset by Ryburn Publishing Services, Keele University, Staffordshire

Printed in Great Britain by
The Amadeus Press Ltd, Huddersfield, West Yorkshire

Contents

Introduction

TVR is a charismatic marque which has survived against all the odds for many years. Not only that, but it seems to have prospered by breaking all the rules. How many other car-makers can be named which have gone broke so often, yet approached their 50th anniversary with a great reputation? How many other car-makers have stayed independent for so long? How many others have resisted the siren call to massive expansion, to building huge new factories on trading estates and to believing in their own publicity? And how many could have achieved so much while basing themselves in an industrial backwater – Blackpool?

The first time I set out to write a book about TVR, I intended it to be no more than a factual, blow-by-blow account of the way the TVR motor car developed from the first machines built in the Fifties. All that, I hope, was included in those pages, but an additional and exciting element repeatedly crept in, and in the most unexpected places. This was the 'character' of the TVR, a quality with which all the cars have been liberally endowed.

Writing my first TVR book converted me totally to the cheerful 'can-do' philosophy of building cars, which the little Blackpool-based company has always embraced. When the chance came to extend my knowledge of these splendid sportscars, I drove eagerly to Blackpool to start again. The original book covered all models from the original Grantura to the Tasmins of the Eighties, but since the early Eighties there was such an explosion of new TVR models that a further update was impractical. Fortunately Motor Racing Publications *and* TVR both encouraged me to dig deeper, to turn one book into two new and more detailed volumes.

All three of us – manufacturer, publisher *and* author – agreed that the tidiest way to tell this story was by splitting the narrative at the end of 1979. In every practical way, therefore, Volume 1 is a complete story of the early period.

The companion volume, on the other hand, picks up the TVR story in 1980, when the Tasmin family was born, and brings it up to date.

This *Collector's Guide* is not meant to be a high-flown, romantic, revs-stunts-and-tall-stories account of the cars from Blackpool. Instead it sets out to provide the mass of statistics and identifying material that every true enthusiast needs to have when he is looking for a new car, or wants to restore his TVR to its original condition. The acknowledgements follow this introduction, but it should freely be confessed, right now, that the job could not possibly have been accomplished without unlimited help from Martin Lilley and all his staff at Blackpool.

The first TVR went into production at about the same time that I first became involved with the motor industry, so it could be said that we have both grown up in it and gained our experience together. I would be the first to admit, though, that the life and times of TVR has been infinitely more absorbing and adventurous than mine! However, the coincidence of dates and my gradual conversion to a study of the industry's heritage has let me keep a constant watch on the marque's progress.

It is a miracle that TVR still exists, for if all the laws of probability and finance had consistently applied to the marque it would have died some years ago. The original Granturas were crude machines built by a hilariously inexperienced team. For years, the marque stumbled from crisis to crisis, from boss to boss and from one set of capital to the next. TVR Engineering was displaced by Layton Sports Cars, TVR Cars took over from that and Grantura Engineering kept the pot sporadically boiling until it slid into bankruptcy in 1965. It needed Arthur and Martin Lilley – father and son – to make the bold move of buying up the assets, such as they were, breathe life into the corpse, set up TVR Engineering Ltd and bring about the revival which has continued, not without problems, to the present day.

Before the Lilley influence took hold, there had only truly been one type of TVR which, one chassis change and several engine options later, was still a rather unreliable machine. It was only after stability set in, towards the end of the Sixties, that the really remarkable advances in engineering, performance and product quality began to occur. It was all part of a complete change of image. Before Lilley, there was always the feeling in TVR-watchers' minds that the firm might collapse; after Lilley, as it were, that feeling was replaced by one of permanence.

The character of the TVR has always shone through, and this is nowhere more obvious than in its engineering. Each and every TVR has had a multi-tubular chassis-frame and all-independent suspension; each and every TVR has had a glassfibre bodyshell. Every car, too, whether endowed with a powerful engine or not, has undeniably been a sportscar, and TVR could never be accused of providing boring machines, even if they were not all very fast.

Less than 5,000 glassfibre-bodied TVRs were produced between 1958 and 1979, but because of their rot-resistant construction many of them have survived. Because bodyshell restoration is so much more practical than with steel or aluminium types, cars are continually being reborn. Somewhere in these pages – whether in the form of a Griffith or a Vixen, a Turbo Convertible or even a Ford-engined Grantura, there may be a TVR for you. But don't all rush at once – the supply is limited!

GRAHAM ROBSON

Acknowledgements

When I started researching material for this much revised book, I thought that fact-gathering for what is, in effect, a 35-year history would be straightforward enough. I was wrong. It needed a great deal of effort and a lot of assistance before I was satisfied that all available knowledge gaps had been filled.

My job, therefore, has been made much easier by the following experts:

For encouragement, for their personal efforts with the archives, for raking back into their memories, and for reading the manuscript progressively as it took shape: Martin Lilley, Stewart Halstead, Carole Newton, Stan Kilcoyne, Mike Penny, Noel Palmer and James Pillar, of TVR Engineering Ltd.

For their enthusiasm in maintaining interest in the marque through thick and thin: The TVR Car Club, and particularly two secretaries consulted over the years, Douglas Manuel and Roger Cook.

For his personal involvement, in business and pleasure, with TVR, and for his immense stock of historical records and illustrations: John Bailie and his staff at Image Publicity in Blackpool.

Apart from TVR Engineering Ltd and Image Publicity, for providing photographs: *Autocar*, *Thoroughbred & Classic Cars*, *Motor*, Gerald Jinks, Mirco Decet and Douglas Manuel.

Those recorders of the motoring scene, for their archival importance: *Autocar, Motor, Autosport, Thoroughbred & Classic Cars* and *Road & Track*.

My personal American connection: Richard Langworth, of Dragonwyck Publishing Inc, New Hampshire, USA.

Lastly, of course, I should acknowledge yet again the debt which all TVR enthusiasts owe to Arthur and Martin Lilley for their faith in the marque. If they had not bought up the thrice bankrupt remains of TVR in 1965 it would almost certainly have died there and then, and this story would never have had to be written. Even since then, there have been as many hard knocks as there have been easy successes, and I admire them all the more for that. Through their persistence, I am sure that TVR will prosper for many years to come.

October 1993 GRAHAM ROBSON

By 1956, two years after the first TVR had been sold, this predecessor of the Grantura had evolved. Beneath the skin are the chassis, suspensions and Coventry Climax engine first seen in Trevor Wilkinson's Jomar design, but the bodyshell was fashioned by TVR. Jack Pickard, one of Wilkinson's early associates, is at the wheel of this car, which was intended for competition use, hence the closed-in passenger space and the combined headrest/fairing. (*John Bailie*)

CHAPTER 1

Ancestors and heritage

Prototypes and specials

The story of TVR effectively begins in 1947, for it was in that year that a young man called Trevor Wilkinson built his first 'special'. However, the first-ever car which Wilkinson dubbed a TVR did not follow until 1949, and the first TVRs built in series – a very limited series – were not put on the market until 1954.

Trevor Wilkinson was born in Blackpool in 1923 and left school in 1937, at 14 years of age. He then joined a local garage to serve his apprenticeship as a mechanic and stayed in Blackpool throughout the war years, helping in the family business (a shop selling baby's prams) for some time. In 1946, with no more experience and mechanical training than he had picked up in his teens in the motor trade, the young Wilkinson determined to set up business on his own. At the age of 23, therefore, and in spite of some parental opposition, he acquired an old wheelwright's shop in Blackpool and established Trevcar Motors. Not only did the tiny business take on car repairs, along with the buying and selling of cars, but it became involved in a variety of light engineering enterprises.

In 1947, Wilkinson owned an old Alvis Firebird – a four-cylinder car of no particular engineering merit which had been built in Coventry during the Thirties – and it was on this basis that he built up his first 'special'. The rolling chassis was not altered, though the standard body (which presumably was on its last legs anyway) was discarded, and a new light-alloy two-seater sportscar was built to take its place.

Later that year, TVR Engineering replaced Trevcar Motors, Wilkinson was joined by Jack Pickard, and their collective enthusiasm turned to the idea of building a sportscar which would be 'special' from the ground up. Giving it a name – the initials TVR came by shortening TreVoR's christian name – was easy enough, but designing and building the car proved a challenge. The original TVR was not on the road until 1949 and the third car in this embryo series was not completed until 1951.

It is important at this point to remember that neither Trevor Wilkinson nor Jack Pickard had any formal engineering training. Both had left school at 14 and neither had had any experience or knowledge of the way in which cars were normally designed. Factors such as weight distribution, spring rates and damper settings, and costing for series production were completely foreign to them. They knew no more than they had learned from looking at other cars – some which they liked, some which they did not – and from reading about them in the motoring publications of the day.

In 1949, therefore, with a great deal of enthusiasm but little money, Wilkinson set out to design and build his first car, which was based around a multi-tubular chassis, of which the upper tubes also supported the body skin panels. There was trailing-link independent front suspension, coil springs from the crash bumpers of fairground bumper cars, a Morris Eight rear axle and a side-valve 1,172cc Ford 100E engine and its matching three-speed gearbox. Except for the tubular chassis and the skin panels, every component was second-hand. The panel-beater who had built the Alvis Firebird's body, Les Dale, produced a two-seater 'torpedo'-style shell for the new car, which used Morris Eight Series E road

The first TVR officially to be put on sale was the multi-tube-framed car introduced in 1954. This version, with an RGS Atalanta glassfibre bodyshell, was known as the TVR Sports Saloon. The chassis used many Austin A40 Somerset parts, and Trevor Wilkinson originally intended them to be fitted with A40 1,200cc engines. However, about 20 chassis were supplied between 1954 and 1956 with a variety of engines, the most powerful being the 2½-litre twin-high-camshaft Lea-Francis unit. (*John Bailie*)

wheels and had cycle-type wings.

Having built the car, Wilkinson then sold it to his cousin for £325 and immediately set about building another. The second car was very similar to the first, except that it was treated to coil-spring-and-wishbone independent front suspension from the current Austin A40 Devon saloon, and the third car, which followed in 1951, not only had the A40's front suspension, but also that model's 1.2-litre pushrod ohv engine, four-speed gearbox and spiral-bevel back axle; rear suspension, as before, was by half-elliptic leaf springs, with no extra location. The third car was used for a time in local competitions, driven by Wilkinson, and it won several awards.

By now, the sportscar 'bug' had well and truly taken hold of Wilkinson and TVR Engineering. In 1953, Trevor settled down to design a new car – not merely a 'special', but one which he intended to sell in a series. Although it was obviously and directly descended from the layout of the

One of the first-ever multi-tubular backbone-chassis designs of 1955, fitted with a light-alloy Coventry Climax overhead-camshaft engine, and destined for Ray Saidel in Manchester, New Hampshire, in the USA. This was the first TVR to have independent front and rear suspensions, and Saidel bodied them and called them Jomars. This was the true ancestor of the Grantura. (*John Bailie*)

Front view of the 1955-style backbone-chassis layout produced by TVR for sale to Ray Saidel. This was intended to be bodied as a Jomar sports-racing car and had a Coventry Climax engine. Both front and rear suspension links are those used at the front of the VW Beetle. Note the low position of the water-cooling radiator. (*John Bailie*)

original three cars, the new TVR, which went on sale in 1954, was different in almost every detail.

As did so many other aspirant car manufacturers of the Fifties, Wilkinson decided to offer his cars as complete and built-up machines, ready-to-assemble kits or even little more than chassis-frames. In other words, he was quite happy to supply exactly what his customers demanded, for it was in this way that he hoped to expand. In particular, he chose to source as many as possible of the mechanical components from one mass-produced model, and he decided that it made no economic sense to build his own light-alloy bodies. The car he selected to provide many mechanical parts was the Austin A40 (by now, incidentally, rebodied and called the Somerset) and for bodies he resolved to buy in whatever he needed.

The 1954 variety of TVR, therefore, was based on a new design of multi-tubular chassis-frame (something like that of the 'X' type of Allards), with upper and lower tubular longitudinals and vertical, horizontal or diagonal cross-bracings where appropriate. It was by no means a pure spaceframe (Colin Chapman, no doubt, would have laughed very loudly if he had bothered to analyze the stresses in the structure) but Wilkinson had never intended that it should be. Instead, it was solid, simple and neatly made. The independent coil-sprung front suspension, wheels and spiral-bevel back axle were all from the Austin A40 (and were bought from the local BMC dealer in Blackpool), while the rack-and-pinion steering was from a Morris Minor. Most of the chassis tubes were of 1.5in diameter, with 16-gauge walls – a cross-section to be used by TVR models built until the present day and to be found even in the Tasmin of the Eighties.

Because of the way in which the car was sold, there was really no such thing as a standard engine, although Wilkinson originally intended that the 1,200cc Austin A40 engine and gearbox should be just that. In rather more than two years and about 20 cars or chassis, A40, Ford 100E, MG TF 1500, MGA or even 2½-litre Lea Francis engines and transmissions were used.

For their bodies, TVR went to firms already catering for such a demand. The sleek fastback coupe styles, now known as TVR Sports Saloons, had glassfibre bodyshells by RGS Atalanta, while other open shells were provided by Rochdale and Microplas. The Sports Saloons were advertised for sale at £650 (plus Purchase Tax if completely assembled), though only three were delivered.

While this sporadic activity was taking place, and TVR Engineering was still paying its way in Blackpool with engineering work connected with the vast entertainment industry which dominated the sea front between Lytham St Anne's and Fleetwood, Trevor Wilkinson was already making plans for yet another type of car, this time more advanced, sportscar-like and specialized than before. To TVR enthusiasts, it was important for two main reasons: it had a backbone-style frame and the first TVR-built glassfibre bodyshell. It was also the first TVR to be built in the company's 'new' premises.

In the meantime, Wilkinson had been amazed to receive an order for a rolling chassis from one Ray Saidel, of

By 1957, the Grantura theme was becoming very evident. This was one of the original TVR Coupes – the first closed car to be designed and built at Blackpool. Like earlier cars, it retained the Wilkinson-designed backbone-style chassis-frame, with all-independent trailing-arm (VW Beetle components) suspension. Already evident are the Ford Consul windscreen, the tiny doors and the typical TVR Grantura proportions... (*John Bailie*)

...but the rear-end is still a unique and messy layout. It has an exposed spare wheel, but it does at least have a separated 'openable' luggage compartment. No TVR built from 1958 until 1976 would ever be so convenient again. The registration number originated in Blackpool. (*John Bailie*)

Manchester, New Hampshire, USA, which according to the order should be fitted with a Coventry Climax FWA light-alloy engine of the type which was already making its name as a sportscar racing unit in the UK and Europe. It was to be the start of a fruitful, if turbulent, relationship.

Wilkinson's aim with his new chassis design was to lower the seats and, therefore, the entire car. This would improve both the aerodynamics and the roadholding. With the existing chassis, the height of the seats was determined by the fact that they had to sit on top of the floor of the bought-in bodies, which were themselves perched on the tubular frame.

The new chassis, therefore, was laid out as a basic backbone type (though it is doubtful if Wilkinson called it anything as scientific as that, at least, not at first). Instead of having widely spaced longitudinal members, its structure was based around an open box-section of four tubular longitudinal members, which were swept closer together (and the top tubes lowered) to provide an acceptable section through the cockpit. The engine and transmission, therefore, were completely wrapped around with steel tubes, suitably braced

and reinforced to increase the beam and torsional properties, and the frame itself started and finished at the front and rear suspension mountings. There were outrigging tubes, of course, along the level of the straight bottom tubes, and the result was that the seats could be mounted right down at floor level, no more than about six inches from ground level.

This, too, was the first TVR with independent front *and* rear suspension. Once again, Wilkinson was anxious to use as many as possible of the chassis components from the same model and since there was no British car with the sort of layout he needed, he turned to foreign models. For the new frame, and for the design which was to be used on all TVRs built between 1957 and 1962, he chose trailing-link front and rear suspension, sprung on transverse torsion bars, and he took his assemblies from the VW Beetle. Note that front *and* rear TVR suspension used the trailing links and transverse torsion bar of the Beetle's front suspension, for the Beetle's rear-end was a swing-axle layout to match its rear-mounted engine.

The rolling chassis for Ray Saidel was soon given a USA-built bodyshell and named a Jomar (after JOhn and MARgaret, Saidel's two children), and more orders for this all-independent chassis followed. Out of this design came the original (pre-Grantura) TVR Coupe of which, illogically enough, three closed and three open examples were to be built. Unlike the original Saidel car, which had VW-type wheels to suit the VW brakes and suspension, the Coupes had 11in Girling brakes and Dunlop 48-spoke wire wheels, but all were fitted with Coventry Climax engines, backed by MGA gearboxes.

The body style of this car evolved by the most classic cut-and-shut methods. Once thought to be from a Rochdale shell, it is now known to be from that of another kit-car shell, the Microplas Mistral. It did not at first look like what was to become the 'classic' TVR coupe style, as a notchback roof on rather conventional lines was chosen. However, gradually and inexorably, these hand-to-mouth projects were all leading towards a definitive TVR design, for by 1957 certain of the basic dimensions had already settled into place. The 1957 Coupes, like the Granturas which were to follow, had established the backbone frame, all-independent trailing-arm suspensions, VW worm-and-sector steering, a 7ft wheelbase, 4ft 4in front and rear tracks, and the classic early-Grantura overall dimensions of 11ft 6in length, 5ft 4in width and 4ft height.

In the meantime, at the beginning of 1956, Wilkinson had sold the original Trevcar/TVR Engineering building at Beverley Grove, Blackpool, and taken a long lease on part of a redundant brickworks on the Hoo Hill Industrial Estate, at Layton, also in the outskirts of Blackpool. By this time, too, Bernard Williams had become a director of the company, and extra finance had been provided by Fred Thomas, of Bolton. It was the start of an expansion in business activity which would eventually prove to be fatal to the original concern. The final nudge, from the building of cars in tiny quantities to the building of TVRs in something like series production, followed in 1957, when Ray Saidel exhibited a car at the 1957 New York auto show (having been appointed the first-ever TVR/Jomar distributor) and was astonished to receive a flood of orders, few of which the company could possibly fulfil.

Working on ideas suggested to them by Saidel (one of which was that he would like to be able to offer a sleeker, fastback model), and with an idea of shaking down the backbone frame design into more of a series-production model, Wilkinson and Jack Pickard set to work on what was to become the first true production TVR. For 1958, therefore, a new era at TVR was beginning. With Wilkinson as Managing Director and technical chief, Jack Pickard at his elbow, Bernard Williams effectively heading the sales effort and Fred Thomas providing finance, TVR Engineering faced up to an exciting future. No-one could possibly have guessed that there would be a complete financial reconstruction within 12 months, a bankruptcy within five years and a further bankruptcy only three years after that. For the founding fathers of TVR, it was going to be an exciting, if bumpy, ride.

CHAPTER 2

Grantura Marks I, II and IIa

1958 to 1962

The series-production TVR Coupe, which was not christened 'Grantura' at that stage, but retrospectively became known as such, was revealed in 1958, at a time when only one representative car had been completed. *The Autocar* and *The Motor* made mention of the launch and clearly the two long-established weekly magazines attended the same press briefing, for they both emphasized that the launch was the successful culmination of a couple of years' development, testing and racing (by Ray Saidel and others in North America), that Trevor Wilkinson and Bernard Williams were the two most important individuals in the Blackpool company's team and that a choice of alternative engines would be made available. Interestingly enough, at this point the prices quoted were for complete cars, though TVR always made it clear that they were prepared to supply kits, indeed that they would supply cars to which the customers could fit their own engines if they wished.

Their launch got off to a flying start because the prototype was displayed for a while in the Manchester showroom of H & J Quick Ltd, the prestigious Ford dealers, at which time a white lie was told to the effect that: 'So far, a batch of 10 has been produced ...' when, as already made clear, the only complete TVR was the one on display in the showroom in Manchester.

The Autocar set the seal of respectability on the Grantura (which had received its official title by then) by becoming the first magazine to publish a detailed technical analysis of the car. On March 6, 1959, not only did they devote three pages to the Grantura, but they graced it with one of their famous cutaway drawings to illustrate the complexities of the chassis.

The Mark I, like all Granturas built until mid-1962, was based on the multi-tubular chassis-frame which Trevor Wilkinson had begun to design in 1955, and which had first been supplied (to Ray Saidel, for use as a Jomar) in 1956. The engine bay, enclosed by the chassis backbone tubes, the front suspension crossmembers and the glassfibre bulkhead, was roomy enough to accept several different engines. At first there was really no such thing as a 'standard' engine, though *The Autocar*'s description was of a car equipped with a 1,216cc Coventry Climax FWE engine (with an MGA gearbox). Other engines offered, and fitted from time to time, were the obsolescent side-valve 1,172cc Ford 100E unit, which was matched by the three-speed gearbox found in the Anglia and Prefect saloons; a supercharged (by Shorrock) version of this; the new overhead-valve, over-square, 997cc Ford 105E engine and four-speed gearbox from the 1960-model Ford Anglia; and the 1,489cc MGA engine and four-speed gearbox.

The kernel of the Grantura's design was the use of VW trailing-link independent suspension at front *and* rear. Twin parallel trailing links were employed at each corner of the car, joined – top and bottom – by laminated torsion bars enclosed in the transverse frame crossmembers. The bars were anchored at their centre by bolts screwed into the centre of the tubes, and were effectively short stiff bars mounted end-to-end in the same tube. Incidentally, although this was a VW system through and through, the patriotic *Autocar*, in January 1958, had chosen to describe the trailing arms as 'recalling the Atalanta system', and the torsion bars as being like those

Side view of one of the early series-production Grantura Mark I models, showing the stubby and completely characteristic lines, the wrap-around rear window, the smoothly contoured tail (contrasting with the lumpy outline of the 1959 Coupe) and the lack of opening quarter-windows in the doors. However, modifications have already been made to the original, for this particular car has exposed headlamps with chrome decorating rings. (*John Bailie*)

of the defunct Daimler Conquest/Lanchester Leda models!

The suspension on the TVR was so stiff that no anti-roll bars were needed and the telescopic dampers must have had quite an easy time. The geometry meant that the wheels rolled at the same rate as the chassis and bodyshell, which is to say that the effectiveness of the tyres decreased as the roll increased. With such hard suspension, however, there was undoubtedly very little roll, even at high cornering speeds. The rest of the front suspension and steering was an ingenious amalgam of standard items – something for which TVR became noted as the years progressed – for although the kingpin was a light-alloy item manufactured by TVR, the stub-axle was from a BMC (Austin A55) saloon, the brakes were as used on the current Austin-Healey 100-Six and the steering box was a Ford item. At the rear, the same basic kingpins were used, but the toe-in was controlled by the fitment of extra radius-arms, which were really track rods fixed securely to the chassis at the inner ends; these could be adjusted for length in the normal way, an operation made necessary by the fitting of Metalastik bushes in the kingpins and rubber bushes at the track rod end.

Girling 11in diameter drum brakes were standard (virtually the same as those fitted to the Austin-Healey 100-Six, but not to the Austin-Healey 3000, which had discs), and bi-metal Alfin drums were specified, as were Dunlop centre-lock wire-spoke wheels, also of Austin-Healey type. Even in 1959, when *The Autocar*'s analysis was published, it was stated that Girling front discs would soon replace the drums and that ventilated disc wheels would soon be standardized (with a £35 price reduction). In the event, the disc wheels were never employed and the use of disc brakes was delayed until 1961, and the advent of the Grantura Mark IIA.

Other mechanical fittings included a crossflow water radiator, mounted well forward and low down in the nose, an 8¾-gallon fuel tank (TVR-manufactured) centrally mounted behind the line of the rear suspension/final-drive unit, and the spare wheel tucked well down in the tail, behind the tank, covered by interior trim. No matter which engine/gearbox combination was chosen, the Grantura's chassis-mounted differential/final-drive was the BMC B-Series hypoid-bevel

Parked in typically industrial North Country surroundings, this early Grantura Mark I shows off its very smooth front-end styling, which had never been near a wind-tunnel, but was demonstrably efficient and wind-cheating. One tiny badge on the nose was all that advertised the TVR make. The wipers have yet to be fitted to this car. Note the huge expanse of bonnet combined with the front wings, which was front-hinged and swung up to provide unrivalled access to the engine and front suspension. (*John Bailie*)

This 1958–59 Grantura Mark I is waiting to be shipped to the United States, though it has been built in right-hand-drive form. Obvious in this shot are features so typical of TVRs for many years to come, including the panoramic rear window, the smooth lines, the very short rear overhang and the lack of access to the rear stowage space. Somewhere in there was the spare wheel, which had to be extracted through the passenger compartment and one of the doors! The quarter-bumpers were more for visual effect than protection. (*John Bailie*)

design (as fitted, from the mid-Fifties, to cars as diverse as the MGA and the Austin-Morris Farina-styled saloons), with a variety of ratios which depended on the revving capacity of the engine chosen, but the case was a light-alloy TVR design.

However, it was the Grantura's looks which made the initial impression on a potential customer. TVR might have started from the unlikely base of Rochdale styling and used hit-and-miss, cut-and-fill methods of arriving at the final shape, but the end result was undeniably distinctive, even if the designers would have liked it to look longer and even more wind-cheating. It was, in fact, the basic shape from which all TVRs built until the end of the Seventies would evolve and it stood the passage of time remarkably well.

The choice of glassfibre production was almost inevitable, for no British manufacturer could produce hand-built metal shells for anything except an expensive car, and at the rate of production forecast by TVR there was no chance of having pressed panels. It was a simple fastback shell, with no access to the stowage space from outside the car (which made removal of the spare wheel a tricky and time-wasting process – not to mention hard work), but with a full-width bonnet moulding hinged at the front of the chassis-frame and giving almost ideal access to the engine bay and front suspension.

The windscreen was that of a Ford Consul/Zephyr Mark II – available in either toughened or laminated glass – while the large curved rear window was Perspex, and special to TVR. On the first cars, the lines were extremely smooth, as the headlamps were faired-in (like those of the 1957 Coupes), and there were no quarter-windows in the doors. One considerable disadvantage, which was not to be alleviated for some years, was that the doors were extremely short, which made access to the seats rather difficult, though the driving position itself was very comfortable.

Production of tubular chassis-frames at the Hoo Hill works in Blackpool. These four frames represent up to two weeks' complete production of TVRs, for output at the end of the Fifties was still extremely limited. Many changes were made to this design before it gave way to the coil-sprung Mark III frame of 1962. (*John Bailie*)

In the first year, TVR found it difficult to produce the Granturas at the rate which Ray Saidel demanded. After the phenomenal interest shown in the prototype car at the 1957 New York auto show, Saidel was convinced that he could sell just as many Granturas as TVR could deliver. Their problem, however, was not only that they were still learning how to build glassfibre coupe bodies in quantity *and* to the appropriate level of constructional quality, but they were also finding the usual financial problems connected with the rapid expansion of a small business.

By the end of 1958, with less than two dozen cars built in all, TVR Engineering found themselves with plenty of orders, but with a large and growing cash-flow crisis. The result was that TVR Engineering was wound up in December 1958 and a new company, Layton Sports Cars Ltd, was formed to take over the business. The car itself, the premises in which it was built and the principals (Wilkinson and Williams) were not

A detail of the rear suspension and driveshaft layout of the early-type trailing-arm TVR chassis. The trailing arms and the transverse torsion-bars are from a VW Beetle front suspension, though the driveshafts are special, as is the cast-alloy rear-suspension upright. (*John Bailie*)

From the rear, the original type of TVR chassis can be seen to be very low and squat. The final-drive housing was special for this car, but the rear suspension was from the front of a VW Beetle. This chassis would eventually have gone to the United States to become a racing Jomar. (*John Bailie*)

changed, but there was a change in financial backing. Shortly after this, early in 1959, another company, Grantura Engineering Ltd, was also formed. In a very complex legal and financial situation, they took over much of the purchase and manufacturing responsibilities from Layton Sports Cars, though the two companies operated from the same premises; Layton, indeed, was the 'front' organization which took all the publicity.

Early in 1959, when *The Autocar* published its technical analysis, the car had changed significantly in detail, and it was stated that: 'Of the 30 or so TVRs so far made, some two dozen are in America', and it was also made clear that the American cars were being sold as Jomars. An early example of TVR optimism was evident, for the factory had assured *The Autocar* that: 'Present production rate is about six cars a month, but the rate is rising steadily and is planned to reach

The ancient but sturdy design of side-valve Ford engine – the 1,172cc Type 100E unit – which powered some of the original Grantura Mark Is. This was Ford's own installation, and for TVR use a different gear-change was adopted. Some 100E engines were supercharged, but it was still not the most powerful and satisfactory way to power a TVR – the Coventry Climax engine was best for that. (*Ford Motor Company*)

about five a week by the end of this year'; what actually happened is that production slowed down during 1959 as the outstanding orders either disappeared or tailed off.

In the meantime, the headlamps had come out into the open, swivelling quarter-windows had been added to the doors, which made the drop glasses even narrower than before, and flares had been added above the front wheel-arches and wings, the flares actually being part of the lift-up bonnet moulding. Twin 'nostril' vents were to be found in the bonnet on some cars, above the main radiator air intake, and during 1959 small vents were added behind the front wheelarches, which helped to channel hot air out of the engine bay (particularly on the most powerful Climax-engined cars).

At this stage in the development of the TVR production car, the specification was often altered to suit the individual customer's requirements. The fact, therefore, that bonnet air vents were only supposed to be on Ford-engined cars at first

Two views of the 1,588cc BMC MGA 1600 engine and transmission which was used to power many of the Grantura Marks I, II and IIA built between 1958 and 1962. Early examples had 1,489cc engines, which were slightly less powerful and had minor visual differences. Most of the lengthy transmission casing houses the output shaft and the complex but precise gear-change linkage. Some Granturas were treated to the HRG-Derrington conversion of this engine, which included a light-alloy crossflow cylinder head. (*BL*)

Many older TVRs on the market will inevitably have been 'customized', rather like this Mark I Grantura, which features bonnet straps, wing mirrors and the engine bay air vents and front wing flares which had been added to the specification in 1959. This car also has the door quarter-windows as fitted to all but the first cars and the rather more substantial bumpers which soon became a feature of the design. (*John Bailie*)

didn't mean that they were never found on cars with other engines. It was also possible for a customer to order a lightweight bodyshell, so that he could race the car, though this, like all the others, was bonded to the multi-tube frame on assembly, which made things awkward (and more expensive) when crash repairs became necessary.

Among the special fittings found on early Granturas (many of which may now have been taken out of surviving cars) were the Stage 2 or even Stage 3 tuning kits on Climax-engined cars. Some were of 1,098cc, to race in the 1,100cc classes, and some were the 1,216cc (FWE) units found in all 1959–63 Lotus Elites and several other specialist sportscars of the period. The KSK four-speed gearbox conversion for Ford 100E engine/gearbox models could also be ordered (*The Autocar*'s 1959 description mentions this), and by 1960 it was possible to ask for Girling front wheel disc brakes at extra cost as well.

There is only limited scope in a *Collector's Guide* for discussion of the commercial events which affected the evolution of a marque or a particular model. However, it should be pointed out that Layton Sports Cars' financial affairs often seemed to be in disarray, with production *and* orders fluctuating considerably from season to season. In 1959, this resulted in Ray Saidel relinquishing his TVR dealership in North America, which therefore meant the death of the Jomar and the loss of the US market to TVR for some time. Several individuals (including financier Fred Thomas) moved out, but an ex-Rolls-Royce engineer called John Thurner joined the company as its new technical chief, taking over from the car's originator, Trevor Wilkinson, who was gradually pushed further into the background.

However, although only 54 cars were built in the 18 months leading up to the end of 1959, the total production run of TVR Granturas was brought up to exactly 100 by mid-1960, at which time the rate of production was increasing markedly and the Mark I was superseded by the Mark II.

The Grantura Mark II was effectively a lightly modified Mark I, with several improvements and a little visual retouching. The rolling chassis was basically unchanged and the bodyshell itself was virtually untouched. This time it was

This shot, taken in 1976 outside the TVR factory at Bristol Avenue, gives a vivid and revealing comparison between the TVR 3000M Turbo model and a 1961 Grantura Mark IIA. The family resemblance is clear, though it is quite obvious that there is no direct link between the two designs. The Mark IIA has been modified by its owner, for the single bonnet air intake was never part of the standard specification. (*TVR*)

Arnold Burton competing in the 1961 Tulip Rally, high up in the mountains of France, in his Mark II Grantura. He finished, but only just, as he was classified last of all. (*John Bailie*)

the turn of *The Motor* to describe the latest car (in their issue of July 6, 1960), when the realignment of mechanical offerings and the styling changes were clarified.

The principal engine was now the MGA unit, a few early cars having the now-obsolete 1,489cc capacity, but most having the 1,588cc MGA 1600 capacity which, in 80bhp form, had been used on MGAs built from mid-1959. The standard options were the 997cc Ford 105E unit, or the Type FWE 1,216cc Coventry Climax engine.

Visually, the flared front wheelarches of the later Mark Is were now matched by flares over the rear wheel cut-outs and the rear wings had been reshaped and extended (almost becoming small fins in the process) and were treated to different (standard Lucas) tail-lamps and separate flashing indicators. The same type of amber indicators were also fixed to the bonnet panel, behind and at each corner of the nostril intakes.

All the optional extras developed for the Grantura Mark I were carried forward, except that the side-valve Ford engine was no longer available as it was both underpowered and out of production.

The 1,216cc Coventry Climax Type FWE single-cam engine was a snug fit in the engine bay of the Mark II Grantura. This was *Autocar*'s road-test car of 1961 which, with 83bhp, returned a maximum speed of 101mph. Details clear in this picture include the location of the radiator header tank, which is behind and above the engine, and the prominence of the cross-tube of the chassis-frame, which housed the transverse torsion-bar of the independent front suspension. Tiny bonnet holding-down straps were fitted to this particular car. (*Autocar*)

The Mark II, not only an improved Grantura but a more popular one, achieved higher sales than the Mark I in the first few months, yet it was made obsolete by the Mark IIA of early 1961. The reference to Mark IIA, however, meant very little except that Girling front disc brakes were standardized, the engine line-up was modified yet again and the door quarter-windows could no longer be opened.

In previous TVR books there has been confusion about the availability of the Ford Classic 1,340cc engine. It has been stated that this became available towards the end of Mark II production. However, this cannot be the case as the Classic itself was not revealed until May 1961, the option of the engine was not publicized until September 1961 and by this time the Mark II had been replaced by the Mark IIA.

In the meantime, the standard MGA engine had been enlarged yet again so, allowing for the fact that the first few Mark IIAs were built with 1,588cc MGA 1600 engines, the available engines on Mark IIAs were the MGA 1600 Mark II unit of 1,622cc, the ubiquitous Coventry Climax FWE engine (still standard, of course, in the Lotus Elite), one or two Ford 105E engines and – from the early autumn of 1961 – the 109E 1,340cc Ford Classic engine, which was merely a long-stroke derivative of the 105E unit. However, even on this broad base of 'standard' options, there could be further variation if the customer so desired it. MGA engines could be supplied with the HRG-Derrington light-alloy crossflow cylinder head (which was eventually homologated and was used not only on cars in rallying, but on the 1962 Le Mans cars described in the next chapter), while Coventry Climax-engined cars could be supplied with the ZF gearbox of the type already available with the same engine in the Lotus Elite. Another mechanical oddity was that while the Ford-type worm-and-sector steering was retained for most cars, rack-and-pinion steering was made available for those cars fitted with the Coventry Climax engine. Precise steering, it seems, was only considered essential if the highest-performance model was being ordered...

Meanwhile, further financial and organizational changes had taken place. During 1961, TVR appointed Dick

This picture of *Autocar*'s Mark IIA Grantura road-test car of November 1961 demonstrates the small door aperture which was a feature of all short-wheelbase TVRs built before 1967. This particular car was fitted with a Coventry Climax Type FWE 1,216cc engine. (*Autocar*)

Three-quarter-rear view of *Autocar*'s Grantura Mark IIA road-test car, the first officially tested TVR to exceed 100mph. The bonnet straps (at the tail of the lift-up bonnet moulding) were non-standard. (*Autocar*)

Monnich, of New York, to distribute their cars in North America, while from September of that year, control of Layton Sports Cars was taken over by Brian Hopton and Keith Aitchison, whose Aitchison-Hopton business in Chester was already a leading TVR outlet. Hopton became Chairman and Managing Director, and set about living and working in a style to which all previous TVR directors would have liked to have been accustomed. It was not at all clear, however, whether such conspicuous expenditure would be for the long-term benefit of TVR – or Hopton himself. There was no change in company name, and no immediate change in the TVR production car itself.

The last of the Mark IIAs, however, was eventually to be built in September 1962, some months after its replacement had been announced, raced and stumbled its way into production. Mark IIs and Mark IIAs were in production for just over two years, during which period about 400 cars were sold – by far the fastest selling rate of the TVR so far, and four

Action study of a Coventry Climax-engined Grantura Mark IIA, which shows exactly how hard was the suspension of the early cars. This car is just about on the limit of adhesion, a fact demonstrated by the shape of the outside front tyre. The ground clearance is very limited – a characteristic of almost all early TVRs. (*Autocar*)

times as many sales as that of the Grantura Mark I.

The new Mark III chassis and suspension systems were very different from those used in the original series-production Granturas, so it is now appropriate to survey the life of the car thus far. In rather less than five years, approximately 500 cars were sold with the Wilkinson-designed trailing-arm/torsion-bar chassis, which changed very little during that time. After a while, incidentally, it had become known that the suspension and ride were almost insufferably hard, and it is also worth noting that the ground clearance was somewhat limited (though this altered very little on undulating roads due to the hard suspension). Even when the cars were new and correctly set up, there was some slight degree of 'rear-wheel steering' inherent in the system, due to rubber bushes linking the trailing arms to the wheels, and this tended to worsen as wear took place.

The last of the 1962 Mark IIAs had the same basic bodyshell as the 1958 car, except for a number of detail improvements, not least of which was a growing attention to quality. It has to be admitted that the bumpers (thin on the early Mark Is, rather more substantial on later cars) were little more than decorative appendages, but at least the glassfibre bodyshells could easily be replaced or repaired by any competent glassfibre shop.

In the five years, engines fitted varied from the 35bhp side-valve Ford to a race-tuned MGA 1600 Mark II unit (which could, perhaps, produce more than 110bhp). The maximum speeds varied between the claimed 75mph of those early Ford-engined cars to the 101mph of *The Autocar*'s Stage 2 Coventry Climax-engined machine tested in November 1961, and to well over 120mph in out-and-out racing guise.

However, these early TVRs soon fell out of fashion as the Sixties progressed and it is doubtful if many of them now remain in anything like 'as built' condition. As it was, something like two-thirds to three-quarters of all Marks I, II and IIA were supplied by TVR in kit form, and may not have been in 'as intended' condition by the time the owners had finished personalizing their product. The surviving cars are interesting as early examples of a marque which was to mature and improve progressively in the years which followed.

Yet in 1962, no-one at TVR was thinking much about the old models, as the works was full of preparations for a new car. The combination of Brian Hopton, John Thurner and a new chassis design was about to become obvious.

CHAPTER 3

Mark III, Vixen and others

The new chassis – 1962 to 1972

Although John Thurner had started thinking about a new chassis design for TVR as early as 1960, it was not until 1961 that enough capital could be committed to building a prototype and preparing for production and it was not until the spring of 1962 that the first car was completed and revealed to the public. Even though the Grantura was selling better than ever (although the company was not actually making any money), it was thought that not only did the car need a new chassis – stiffer, more efficient and with a much more supple ride – but also new styling. The new chassis duly came on stream in 1962, though the new styling was never produced for it.

It was Brian Hopton's idea that an official works competition programme would be set up to support the new TVR. Hopton then went out and hired Ken Richardson, who had been at Standard-Triumph for nearly nine years, but had left abruptly soon after the influence of Leyland Motors took hold. Even though the new chassis was not ready, Richardson became TVR's first (and only, as it transpired) competition manager.

The new chassis was meant to eliminate all the basic faults of the original Grantura variety, while remaining faithful to the multi-tubular philosophy and capable of being built on simple jigging at the Hoo Hill works. Although TVR's latest owners wanted the new chassis eventually to be graced by a new glassfibre body, they also insisted that it should be capable of fitment under a lightly modified derivative of the existing body. The only concession they made to the modernization of the Grantura's body – from Mark II to Mark III – was to allow for a slightly longer wheelbase under the same exterior skin. However, although this allowed a little more interior space, and particularly legroom (the seats now being set slightly further back), the original extremely narrow doors were retained.

The most noticeable shortcomings of the 'trailing-arm' chassis, which Trevor Wilkinson had designed and was put into series production in 1958, were that its layout was not technically very efficient, it was not very stiff in torsion and its VW-style suspension gave a very hard ride. Thurner's new chassis set out to rectify all these faults. As it transpired, the new chassis-frame and suspension were even more versatile than had been intended. They remained in production for almost 10 years – from 1962 to the spring of 1972 – and formed the basis of cars with four-cylinder, V6, in-line six-cylinder and V8 engines, and with a range of power outputs from the 63bhp (DIN) of the Spitfire-engined 1300 to the 271bhp (gross) engine fitted to some Griffiths and Tuscan V8s. From model to model, the changes were minimal – here and there a tube or a mounting bracket would be moved, or modified, to allow a different engine, gearbox or final-drive unit to be installed.

The new chassis retained the multi-tubular layout, including the four-tube backbone wrapping around the drivetrain, but it was laid out in a much more stress-efficient and logical manner. Further, the VW-style trailing-link suspension system was discarded completely and replaced by front and rear double-wishbone layouts allied to coil springs - systems, incidentally, still used by many British and European 'Supercars' to this day. Before going on to discuss

The Grantura Mark III's chassis-frame, designed for TVR by John Thurner, was much more rigid than the trailing-arm-suspension chassis which it replaced in 1962. Not only was it more scientifically designed, with a better and more logical attitude to triangulation, but it was much more solid in resistance to bending and to torsional forces. This picture is taken from the rear corner, showing the very sturdy build-up of tubes around the final-drive mountings and the pick-up points for the wishbone rear suspension. Also evident are the forward and rear shut faces for the doors, the tubular outriggers to protect the passengers and the moulded-in glassfibre floor. (*John Bailie*)

the various TVR models which evolved around this new frame, it is necessary to describe the frame in some detail.

As previously mentioned, the basic wheelbase of the 1962 frame was 7ft 1.5in – 1.5in more than that of the obsolete trailing-link frame, and although there were no mechanical constraints except those connected with the existing bodyshell, the front and rear tracks were the same as before.

The design was arranged to make the most of the new wishbone systems, which came first, and the rest of the layout followed from this. At the front, Triumph Herald/Vitesse vertical links and lower wishbones were used, while the upper wishbone was tubular and special to TVR. Special coil spring/damper units, Triumph-type rack-and-pinion steering gear and an anti-roll bar completed a compact and thoroughly modern layout, which was allied to Triumph-type 10.75in Girling disc brakes.

At the rear, there was a special light-alloy vertical link by TVR, allied to wide-based tubular upper and lower wishbones. Due to the fact that the driveshaft had to thread its way out from the chassis-mounted differential unit to the hubs, the combined coil spring/damper unit had to be offset, was pivoted from the bottom of the vertical link and was behind the line of the driveshafts. There was no anti-roll bar at the rear.

Although the frame was built up principally from 16-gauge 1.5in diameter steel tube, like the superseded frame, it was completely different. The obsolete design's principal drawback had been that although there was a straight and continuous run of the lower backbone tubes from front to rear, the engine bay top tube had been stopped abruptly at the bulkhead, welded to a vertical tube, and another much lower tube had then linked this vertical post to the rear suspension mountings. Hence, not only was the frame torsionally weak across the cockpit, but the bending strength, too, had been limited.

The new frame was considerably more sturdy, with tubes running smoothly and consistently (if not absolutely in a straight line) from front to rear. Substantial cages supported the front and rear suspension mountings, and there was a considerable degree of tubular triangulation in most sections. As before, there were substantial outriggers filling in the space between the front and rear wheels, along with vertical sheet-steel panels forming supports for the door cut-outs and shut faces. Although it was by no means a theoretically ideal spaceframe (not even Colin Chapman of Lotus – the acknowledged expert in these matters – could provide such a layout for a two-seater road car with large doors and adequate provision for occupants and the engine) it was a thoroughly practical layout requiring little capital expenditure on tooling, but one which TVR could make in considerable numbers. In practical terms, the only disadvantage of the new layout was that it meant that there was now a very wide and bulky gearbox tunnel running through the cockpit, something which was to be a characteristic of this range of TVRs for the next 10 years.

Externally, there was virtually no indication of the radically new and improved chassis design, for the wheels, tyres and general body lines were unchanged. Even under the bonnet, things were not dramatically different except that the upper longitudinal chassis rails were somewhat lower than before. The only apparent penalty of the new frame was that it was

A 1962 Mark III rolling chassis, with its B-Series MG engine and transmission already installed. Just visible at the rear are the combined coil-spring/damper suspension units and the extra telescopic damper which was mounted ahead of the driveshafts and upper wishbones. (*John Bailie*)

The most important feature of the new chassis design adopted for the Grantura Mark III of 1962 was the use of coil-spring independent suspension. Geometry at front and rear was by double wishbones, and the combined coil-spring/damper units pivoted from lugs on the main crossmember. The engine is a 1,622cc MGA 1600 Mark II unit and rack-and-pinion steering had been standardized. In all respects, the chassis-frame was more sturdy, more advanced and more 'sporting' than that of its ancestors. (*Autocar*)

between 15 and 20lb heavier than the one it replaced, but all the extra weight had gone into making the chassis more rigid and versatile.

To go with the new frame, incidentally, there was a larger fuel tank of 10 instead of 8¾ gallons capacity – still mounted in the tail, behind the final-drive unit.

In the 10 years which followed, very few major changes were made to the design. The wheelbase was increased by 4.5 inches – from 7ft 1.5in to 7ft 6in – in 1967 for the Tuscan V8 SE and in 1968 for the Vixen S2, after which the shorter-wheelbase derivative disappeared completely. Over the years there was a marginal increase in quoted track dimensions, and light-alloy wheels came to displace the steel or centre-lock wires which were normal wear at first.

All in all, 13 different TVR models were marketed in the next 10 years on this one basic chassis design, six with the original 7ft 1.5in wheelbase and seven with the 7ft 6in wheelbase. It would be possible, but confusing, to describe each and every one in the sequence in which they were announced, lived and died away, so this survey will now be split into two sections. A complete family tree of the TVR models featured in this volume is to be found in Appendix A, showing that there were seven models in this chassis series with four-cylinder engines, and six with six-cylinder or eight-cylinder power.

The rest of this chapter, therefore, is devoted to analyzing the four-cylinder models, all of which (with one exception) followed on in sequence from a previous model, without overlap. The exception was the Triumph Spitfire-engined 1300, which came towards the end of the run of this chassis and was built at the same time as the last of the Vixen S3s and S4s.

The Grantura Mark III and Mark III 1800

This was the original model in the 'new-chassis' series, the car conceived by John Thurner for his new Managing Director, and the one intended to take over directly from the existing Grantura Mark IIA, which used the VW-style trailing-link chassis. Although it was originally revealed to the press in April 1962 (and put on show at the New York auto show of that month) and *The Motor* carried a full technical analysis in August, the old-type Grantura Mark IIA remained in production until the autumn, and the TVR factory was really not ready to build Mark IIIs until September/October. In the meantime, the slightly restyled bonnet, with relocated air-inlet grille, and indicators now mounted at each side of it, had also been fitted to the last of the Mark IIA cars, thus helping to confuse historians and those shopping for a TVR.

This explains why the cars raced at Sebring looked like Mark IIs, but were Mark IIAs, something always difficult to understand when the complexities of sporting homologation are considered. The sad story of the 1962 works competition effort will be resumed later in this chapter.

The major decision regarding the new Grantura Mark III was not just that it should have the new chassis, but that the MGA four-cylinder engine and gearbox should be standardized. Although it seems that a few cars were built with what can only be assumed to be existing stocks of Ford 109E 1,340cc and Coventry Climax FWE 1,216cc engines (and their appropriate gearboxes), the time of providing the customer with a large choice of engines was now past. *The Autocar* mentioned the options when they described the car in April 1962, but *The Motor*'s more comprehensive analysis mentioned no option other than that of a tune-up kit centred around the HRG crossflow light-alloy cylinder head. No records of this transition period exist today, but it is doubtful whether more than a handful of Ford or Coventry Climax-engined cars were built.

It should be noted, however, that production of the MGA sportscar ran out in the summer of 1962, before the Grantura

Compared with the original Grantura, there had been many detail changes in the styling of the Grantura Mark III of 1962–64. Notable in this shot are the rear wing extensions, which had been introduced for the Mark II, the flares above the rear wheelarches, which had come along at the same time, and the reprofiling around the rear number-plate area, which was phased-in during the autumn of 1963. This car was one of the first to have the 1,798cc MGB engine. The luggage grid was not standard. (*Autocar*)

One of three Grantura Mark IIIs built for the Ken Richardson-directed attack on the Le Mans race in 1962. Only one car actually started the race, and it retired within minutes due to the chronic overheating of its race-tuned MGA engine, in spite of the fact that three extra air-intakes (said to have been inspired by those of the Ferrari GTO) had been let into the bonnet above the normal Mark III radiator air intake. This picture is taken somewhat later, when the car had been sold to a private owner and was wearing road tyres. (*John Bailie*)

Mark III was properly in production, and it was replaced by the MGB. In practical terms, for TVR, this meant that supplies of the 86bhp 1,622cc engine from BMC dried up almost at once, though they were not forced to make the changeover to 1,798cc MGB engines until the early autumn of 1963. When this happened, there was no increase in weight or bulk, but the power available, in standard form, was increased from 86bhp (nett) to 95bhp (nett) with a similar improvement in the torque, from 97lb.ft to 110lb.ft.

The normal MG gearbox was supplied, and once overdrive became available on MGBs (from the beginning of 1963) that option was also applied to the Granturas; a B-series BMC differential and final-drive assembly was provided in a special TVR light-alloy casing, various ratios being available, to choice. As with the earlier Granturas, there was no separate outside access to the luggage space, and the spare wheel had to be extracted by lifting it out of a recess behind the rear suspension, through the passenger compartment.

Officially, this car was only to be supplied as a kit, so that the customer could avoid paying Purchase Tax (in Britain, that is), though it seems certain that some completely built-up cars were supplied.

In the meantime, however, important personal and commercial events had evolved at Blackpool. In April 1962, sickened by the constant changes in policy and by what he considered to be reckless over-expansion and financial incompetence, Trevor Wilkinson left the company which he had founded. 'His' company, incidentally, had become TVR Cars Ltd following the Aitchison-Hopton takeover, and it was now well on the way to a financial crisis. High living by the directors, and far too much money spent on a competition programme, not to mention the cost of setting up for the new chassis design, all bore down heavily on the concern. Even though the Mark III was proudly displayed at the Earls Court motor show in October 1962, the company was in desperate trouble, and before the end of the month it had closed down altogether.

The Grantura Mark III, therefore, eventually struggled into hand-to-mouth production, basically by employees of Grantura Engineering (which had not collapsed). Messrs Hopton and Aitchison disappeared as rapidly as they had arrived on the scene, Ken Richardson's one-season job (and the competition programme) evaporated and it was left to Bernard Williams and Stan Kilcoyne to pick up the pieces. Before long, too, John Thurner left Blackpool and returned to Rolls-Royce.

The development which saved Grantura Engineering – for a time at least – was the birth and development of the V8-engined Griffith, but consideration of this car belongs to the next chapter. In the meantime, the Mark III went on sale, with an extra telescopic damper to each rear suspension (ahead of the wishbones, effectively balancing the spring/damper unit which was already behind the line of the driveshafts).

In the autumn of 1963, the 1,798cc MGB engine was standardized, minor changes were made to the shape of the tail and the location of the rear number-plate, the front upper wishbone pivots were repositioned to improve the geometry,

Because of the glassfibre construction, it was easy for TVR owners to design and effect their own modifications, such as this opening stowage lid on an early Mark II... (*John Bailie*)

the radiator header tank was positioned above the left-side front wheelarch and the layout of the instruments was revised. At this time, incidentally, a Mark III cost £872 (basic) or £1,054 with British Purchase Tax.

Before the Mark III was displaced by the 1800S, about 90 cars were built, many of them being exported.

The Grantura 1800S

For the 1964 New York auto show, Grantura Engineering had built restyled Griffith models, having reshaped tails, in which the now-familiar 'Manx-tail' shape was prominent, along with a larger and sleeker rear window. (This window, incidentally, remained unchanged from this time until the last of the 3000Ms was built at the end of 1979.) Tail-lights and indicators, instead of being standard round Lucas units, gave place to the three-sector Ford Cortina Mark I (1962–66) combined tail/indicator stoplights. At the same time, the layout of the rear was changed so that the spare wheel was no longer tucked down behind the rear suspension. On the 1800S and subsequent models it was mounted on the flat rear floor of the bodyshell, immediately above the final-drive unit, inside the bodyshell, and neatly covered.

To standardize production facilities, this Manx-tail body style was adopted for the MGB-engined TVRs, which thereupon became 1800S models. S for what? Who knows? S for Short, S for Stubby, S for Sporting? Indeed, does it really matter now?

The first 1800S models were built in the summer of 1964, at a time when the Hoo Hill works was almost completely concerned with building a flood of Griffiths, purely for sale in North America. It was the combination of Griffith problems, financial naivety and sheer commercial inexperience which eventually conspired to bring another builder of TVRs to bankruptcy. The smooth and exciting Trident project certainly didn't help. The sum total of this was that Arnold Burton, who had been a major shareholder of the companies

...which hinged down to reveal the spare wheel, now relatively easy to extract if needed. One wonders why TVR production cars were never treated to this fitting... (*John Bailie*)

since the beginning of the Sixties, refused to fund any more losses. Major Timothy Knott, who had been Grantura's Managing Director for just one year, was pushed out, Bernard Williams pulled out, but kept a controlling interest in a business called Grantura Plastics (which had been set up in 1963 to supply bodyshells to Grantura Engineering – I *said* this was a complex little story!) and Grantura Engineering closed down.

Martin Lilley, who ran the Barnet Motor Co Ltd and had become a TVR dealer in the spring of 1965, was the son of Arthur Lilley, who bought some Grantura Engineering shares just before the crash. Following the close-down, Arthur Lilley made an offer for all the assets, in November 1965, which was accepted. TVR Engineering Ltd, their new company, was founded on November 30, 1965 and before the end of the year the TVR marque was back in business.

At this point, very accurate production quantities, dates and Chassis Numbers may be quoted, as all the TVR Engineering records have been preserved and were made available during preparation of this book, so that whereas before, *approximately* 90 1800S cars were built by Grantura Engineering, there were *exactly* 38 built in 1966 by TVR Engineering.

To get Hoo Hill works back into action, Martin Lilley, appointed Managing Director of TVR (his father, Arthur Lilley, became Chairman), concentrated on building existing models during 1966; new models would have to wait until something approaching financial stability had been achieved. The first TVR to be completed under the aegis of TVR Engineering was a Grantura 1800S, early in February 1966. Of the 38 cars built between February and September, seven had right-hand drive and most had the 4.3:1 final-drive ratio, though a few had the 3.9:1 ratio.

In July 1966, however, after the first 26 Lilley-type 1800S cars had been built, the first of a new derivative – the 'Mark IV' 1800S – was built. Chassis Numbers between 18/027 and 18/044 were intermingled, but thereafter the 'Mark IV's took over completely.

The Grantura 'Mark IV' 1800S

This car was, quite simply, a refined and improved version of the 1800S. The first was built in July 1966 and the last of all in October 1967, at which point it was displaced by the first of the Vixen family. Altogether, 78 cars were built, of which 40 had right-hand drive and 38 left-hand drive.

It was Martin Lilley's policy to upgrade TVR quality considerably, so the difference between the 1800S and the 'Mark IV' 1800S was confined to trim, furnishing and some mechanical improvements. Although *The Motor*, at the time, assured us that the 4.3:1 axle ratio had been standardized, quite a number were built with the 3.9:1 ratio. Very few, incidentally, had disc wheels. The engine/gearbox assembly had been moved forward very slightly, and compared with the 1800S the spring-and-damper settings had been revised to provide a more supple ride, while top *and* bottom wishbones were fabricated by TVR. The most obvious trim change was to the facia, where instruments and controls were now housed on a polished wooden panel. Hidden away was the heater from the MGB (with Rover P5 fresh-air outlets at each end of the facia) and there was a headlamp flasher and self-cancelling indicators.

It is interesting, though perhaps not significant today, that the price of the Grantura rocketed during 1966, from £750 to £998 for the kit. *The Motor*'s December 1966 road-test 1800S cost £1,272 in all, for the heater was listed as an extra at more than £16 and the £40 close-ratio gear option was also fitted.

The Vixen S1

The last 1800S Chassis Number was 18/116 and the first of a new series of cars, the Vixens, followed on immediately – VX/117. Although the Vixen was virtually a Grantura 1800S with an engine transplant, it signalled further advance to stability by the Lilley family, and it sold in greater numbers than the 1800S; between October 1967 and October 1968, 117 S1s were built and sold.

The transplant was that a Ford Cortina GT crossflow engine of 1,599cc, with a peak power of 88bhp (net), mated to its own four-speed all-synchromesh Ford gearbox, took the place of the MGB engine and transmission. The good news for TVR was that the Ford installation was lighter and cheaper than the BMC powertrain, but the bad news for the customer was that the Vixen was marginally less powerful and there was no overdrive option, although the gearbox, at least, now had synchromesh on all forward gears. Like the last of the 1800S models the Vixen had a large, 15-gallon fuel tank.

It was to launch the Vixen, incidentally, that TVR first made an appearance at the prestigious Earls Court motor show. At the beginning of the year, TVR Engineering had taken a stand at the Racing Car Show, and had decided that this really was not their type of scene any more.

Apart from the fact that the Vixen had a noticeably different exhaust note due to the use of the high-revving Ford engine, it could also be recognized by the fitment of the new wide and shallow air intake in the centre of the glassfibre bonnet panel, which fed cool air direct to the air cleaner above the Ford engine's downdraught twin-choke Weber carburettor.

(*Note*: In spite of the fact that Vixen = Ford-power, a very few of the early cars were equipped with an MGB engine and gearbox!)

The Vixen S2

In October 1968, the Vixen S2 took over from the S1, and was a much more practical machine. Mechanically it was virtually unchanged, but it benefited from the use of the longer-wheelbase (7ft 6in) chassis first developed for the Tuscan V8 SE (see Chapter 4), where the extra wheelbase was meant to allow larger doors to be fitted, and for access to and egress from the passenger compartment to be improved.

For the Grantura Mark III, the standard engine was the MG unit. Early examples had the 1,622cc MGA engine, but from the autumn of 1963 the 1,798cc MGB unit was standardized. Here is that MGB engine, neatly installed into the engine bay of a 1964-model Mark III, with the radiator header tank fixed to the wheelarch on the left side of the engine bay. The large-diameter trunking leads from the nose of the car to the heater installation – it might have looked efficient, but it wasn't! Note the neat and simple cooling vents from the engine bay to the outside world, positioned behind the front wheel cut-outs. (*Autocar*)

Compare this side view of the Manx-tail Vixen Series 1 with that of a Mark I Grantura, and the differences which continued to be made to the body style become obvious. The major change from Grantura Mark III to 1800S came during 1964 and involved the reshaping of the tail and rear window. The tiny doors, however, had still not been enlarged, so entry and exit from the passenger compartment was still not very easy. (*John Bailie*)

No more than three cars a week had been built in 1967, but this rose to more than four a week in 1968 and continued to rise as more people warmed to the modern TVR manufacturing philosophy. The Series 2 Vixen, of which 438 cars would be built in almost exactly two years, carried that improvement a stage further. TVR Engineering claimed no fewer than 75 detail improvements over the Series 1 model, but it was the longer wheelbase and the visual retouching which took most attention.

Apart from the longer wheelbase, which made the larger doors so obvious, there were the latest wrap-around Cortina Mark II tail/indicator lamps (instead of the circular Cortina Mark I lamps), a modified bonnet moulding, in which the forward-facing air intake had been discarded and replaced by a long slim bulge, made necessary to clear some engine 'hang-on' equipment, and cars built from the spring of 1969 were given two new depressed-section air intakes, one at each side of the TVR bonnet badge, above and behind the main radiator air intake.

Unseen, but very significant, was the fact that the glassfibre bodyshell was no longer moulded to the multi-tube chassis-frame during manufacture, but was bolted to it. This meant that the entire shell could be lifted off the chassis if required – something very valuable at the time, when crash repairs had to be faced, and these days, when the question of maintenance and restoration has to be considered. Henceforth, all TVR bodyshells would be bolt-on, rather than moulded units.

It is also worth noting that it was during this period that the TVR marque finally achieved stability, TVR Engineering became profitable and steady expansion of production began.

The Vixen S3

The Series 2 Vixen had been a very popular model, but it was replaced in the autumn of 1970 by the Vixen Series 3, at a time when the move from the Hoo Hill works to the Bristol Avenue factory, where TVR production is now concentrated, was in progress. In the meantime, the last of the V8-engined Tuscans had been built, the (British Ford) V6-engined Tuscan had arrived as a 'sister' car to the Vixens, and the short series of wide-bodied Tuscan V8s had also been constructed. All these cars are covered in Chapter 4.

Visually, the Vixen Series 3 model was virtually unchanged, except that the cast-alloy road wheels introduced on the Tuscan V6 were standardized and the front wing vents (behind the front wheels) were now of a more delicately detailed type, and actually were 'Aeroflow' outlet grilles from the rear quarters of the Ford Mark IV Zephyr/Zodiac saloons.

The mechanical improvements were that the 1,599cc Ford engine was now of the rather lustier 86bhp (DIN) Ford Capri 1600GT variety, along with that car's gearbox, but – as before – no overdrive was available.

The Series 3 was built from October 1970 to April 1972 and the last Vixen Series 3 was also the very last chassis of the 1962–72 type to be built. A total of 168 cars were built, 50 in 1970, 97 in 1971 and 21 in 1972.

The Vixen S4

The Series 4 Vixen was one of the 'interim' TVRs built in 1972–73 and was effectively a Vixen Series 3 bodyshell, equipment and powertrain, allied to the M-type chassis which is described fully in Chapter 5. In total, 23 cars were built – 22 in 1972, and just one in 1973. The Vixen series was then discontinued and replaced by the 1600M model.

The 1300

In the entire life of the Mark III chassis-frame, only one four-cylinder model was not a Grantura or a Vixen, and this was the 1300. It was a short-lived and, frankly, unsuccessful car, powered by a 1,296cc Triumph Spitfire engine and that car's all-synchromesh gearbox, and it looked exactly like the Series 3 Vixen and Tuscan V6 models which were current at the time. The car weighed 1,625lb – about 60lb more than the Spitfire itself – and it had considerably more frontal area. Although a TVR 1300 was never road-tested, it may be assumed that its maximum speed was between 90 and 95mph, and that the acceleration would be similar to that of

The badge reveals that this is a Vixen, and the number-plate denotes a car registered between August 1968 and July 1969, while the use of those unmistakable 'Ban the Bomb' tail-lamp clusters from the Ford Cortina Mark I finally clinch the identification – this is a Series I Vixen. (*John Bailie*)

This Series I Vixen is posed at Mallory Park circuit. It is visually non-standard in that the Tuscan V6/Vixen Series 2 type of bonnet has been fitted, the derivative being recognized by the two cold-air intakes in the bonnet, roughly as far back as the centre-line of the front wheels. Many TVRs have been modernized in this way, often due to an accident having necessitated a new section to the body. (*John Bailie*)

the Spitfire, which recorded 0–60mph in 16 seconds and a standing quarter-mile in 20.6 seconds in *Autocar*'s hands.

It was not a success, not only because it was not fast enough (probably *every* production TVR except the Ford 100E and 105E-powered examples was quicker), but also because it was not particularly cheap. The basic price of a 1300 in 1971 was £1,245, which was only £50 cheaper than that of the Vixen Series 3.

Only nine 1300s of the original type were sold, along with five having the M-type chassis (as with the Vixen Series 4 model), while one final 1300 should really have been called a 1300M, as it also had the M-type bodyshell.

The 1962 works competition programme

As mentioned earlier in this chapter, the Aitchison-Hopton regime instituted a works competition programme in 1962,

One consequence of the restyling of the tail, in 1964, to the 'Manx-tail' layout was that the spare wheel henceforth had to be stored under a cover in the stowage space behind the seats. This particular car is a Vixen Series 2, and it demonstrates that this spare wheel position had a devastating effect on the amount of space which could be used for luggage. (*Autocar*)

The neat installation of the Cortina GT 1,599cc engine in the Vixen Series 2, showing the unaltered circular Ford air-cleaner, the trunking to the heater and yet another minor variation of the radiator filler cap position. As ever, access for maintenance was superb. Incidentally, the front wing vents are of an interim type, soon to be ditched in favour of the Ford Mark IV Aeroflow vents. (*Autocar*)

which ended up by costing TVR Cars Ltd a great deal of money and achieved nothing.

To run the programme, Brian Hopton hired Ken Richardson as TVR's competition manager. Richardson had a great deal of experience in racing and rallying road-going sportscars, as he had set up Standard-Triumph's competitions programme in 1954 and had led his department for seven years, during which the Triumph TRs and the prototype twin-cam TR3S and TRS models had raced with some success at Le Mans. However, the facilities available to him at Standard-Triumph were much more lavish than they were to be at Blackpool. Richardson, however, was 'between jobs' – having left Standard-Triumph in the late summer of 1961 – and was anxious to get back into the sporting scene again.

TVR's first foray, in an effort carried out in great haste, was to send a team of three Grantura Mark IIAs to the Sebring

12-hour race. A measure of the rush was that Ken Richardson took up his job in January 1962, the cars left for Florida by sea in February, and the race itself was in March.

The three cars were fitted with the optional lightweight bodyshells and right-hand drive, though they had the new Mark III bonnets and air intakes. Special equipment fitted to the race-prepared 1,588cc MGA engines included the crossflow, light-alloy HRG cylinder head and twin horizontal dual-choke Weber carburettors. There was no success at Sebring, which could surely only have been expected, for one car broke its final-drive, one suffered engine failure after four hours and the third (driven by Mark Donohue and Gerry Sagerman) finished well back after crashing and having to have new steering parts fitted.

The next outing was the International Tulip Rally, which started and finished at Noordwijk-an-See, on the Dutch coast, and ran as far south as Monte Carlo for an overnight halt. There were three works entries – Arnold Burton and Coventry haulage contractor Stan Pateman in their own Grantura Mark IIs, teamed up with Anne Hall/Valerie Domleo in a factory-prepared Grantura Mark IIA. The registration number of Anne's car was 880 ABM, which adds considerably to the mystery as this is a Bedfordshire, not a Lancashire number-plate! Like the Sebring car, this had the new Mark III type of bonnet panel, and it was also fitted with the optional engine parts and the all-synchromesh ZF gearbox.

At one stage in the rally, TVR cars were in the first three positions in their class, but the ladies' car became stuck in third gear for many hours, then Anne Hall put it off the road and retired it. Two Porsche Super 90s then dominated the class, and Arnold Burton took third place behind them.

None of the Sebring race cars returned to Britain, and for the Le Mans 24-hour race, three brand new Mark IIIs were prepared. Incidentally, although production had not properly begun, Brian Hopton was determined to use the Mark IIIs, even though he could not yet deliver replicas, and Richardson was reputedly delighted as the handling with the new chassis was much superior to that of the Mark IIAs used at Sebring. Only one firm entry had been granted, along with one reserve place, so at best only two of the three cars could hope to start.

In the event, only the one car – YFR 751, to be driven by Peter Bolton and Ninian Sanderson – took the start, as not enough other cars dropped out in the practice sessions for the reserve (YFR 752, to have been driven by Rob Slotemaker and Ted Lund) to qualify. In practice, both cars were sorely troubled with engine overheating, which seemed to occur whenever they slowed down considerably after flat-out speeds along the Mulsanne straight. Cooling problems were suspected, but no obvious solution was to hand, and the race car started without modifications.

A padded dash panel was adopted for the Series 2 Vixen, and the switchgear was modified yet again. It was an area which always seemed to be the subject of change, or 'improvement', at TVR – a process which continued throughout the Seventies. (*Autocar*)

The race itself was a disaster. Peter Bolton took the first driving stint, but the engine rapidly boiled its water away, and after only three laps – about 15 minutes of racing – the car stood at its pit, steaming away, the engine ruined.

It was almost the same story at the Tourist Trophy race, held at Goodwood, in August. Three Mark IIIs (the same three cars which had been prepared for Le Mans) were entered, three practised, and there was more engine overheating trouble during practice. Rob Slotemaker's car never left the marshalling area before the start, which left Peter Bolton and Keith Ballisat to take up the running.

The Vixen Series 3 saw one more step along the way towards a smoother and more sophisticated TVR – a process gradually and persistently carried out by Martin Lilley in the late Sixties and early Seventies. From this low viewpoint, the recognition points which distinguish a Series 3 from the displaced Series 2 are the more delicately detailed under-bonnet outlet grilles (from the Aeroflow vents of a Ford Zodiac Mark IV), the optional cast-alloy road wheels and the slightly wrapped-round Ford Cortina Mark II tail-lamp clusters. However, much the most important change was to the wheelbase, which went up from 7ft 1.5in to 7ft 6in at the start of Series 3 production. The extra 4.5 inches went into lengthening the doors, which at last were of a sensible size. (*TVR*)

Bolton's car blew a cylinder head gasket soon after the start, but somehow, miraculously, Keith Ballisat's machine – to his great delight – made it to the finish, 11th overall, and beaten only by Coventry Climax-engined Lotus Elites, which had engines smaller than its own.

That, however, was the end of TVR's works efforts, for TVR Cars Ltd descended into bankruptcy in October, Grantura Engineering took up the challenge in a much more restricted way, and no money was available for works competition activities.

Meanwhile, whereas the four-cylinder-engined cars made up the bread-and-butter of TVR's activities, particularly toward the end of the Sixties and the early Seventies, when production was being pushed up from four cars a week to five, from five to six, and – by 1971 – to seven or even eight, it was the six-cylinder and eight-cylinder TVRs which brought much of the glamour, excitement and even notoriety to the Blackpool concern. Although these were all closely related to the more numerous four-cylinder models, they need a separate study, and this now follows.

CHAPTER 4

Griffith, Tuscan and 2500

Mainly for export – 1963 to 1972

By 1963, the companies building TVRs had already gone through three lots of capital, and the latest concern to take on the challenge – Grantura Engineering – was operating on a very hand-to-mouth basis. There had always been great trouble in establishing profitability, even with the relatively simple four-cylinder-engined TVRs. Nevertheless, Grantura Engineering were then tempted towards more costly and exciting projects when Jack Griffith, an American motor trader who owned a Mark III Grantura, installed a mighty 4.7-litre Ford V8 engine into that car and observed the shattering results.

The story began in 1962, when Gerry Sagerman and Mark Donohue, both of whom had driven a works TVR at Sebring for Ken Richardson, were having their own cars maintained at Griffith's workshops in North America. Donohue's car was a Ford V8-engined AC Cobra and Sagerman's was a 1962 TVR Grantura. One day, just for fun – or so the story goes – mechanics dropped the engine from Donohue's Cobra into the engine bay of Sagerman's TVR, to see if it would fit.

It wouldn't, quite, but Jack Griffith saw the unauthorized 'transplant', was inspired by the very idea of it and determined to do the job properly. He then carried out a complete and operational 4.7-litre installation into a Grantura Mark III, and although there were handling and braking problems to be overcome, he became obsessed with the project and approached TVR on the subject. The result was that TVR's North American distributor, Dick Monnich, brought the prototype over to Blackpool, Arnold Burton put money into Grantura Engineering to finance the idea, and the first deliveries began later that year. Although the car must properly be called a TVR Griffith, Jack Griffith secured the right to market the car in North America, and merely called it a Griffith as he reasoned – quite correctly – that it had been his idea! The car made its public bow at the Boston motor show, in Massachusetts, when it was placarded as a Griffith 200.

This, therefore, was the first of several models designed, if not specifically for North America, then with an eye to exporting most of production to that continent. The Griffiths were followed by the Tuscan V8s, the V8s then spawned off (British Ford) V6 models, and eventually there was also to be a straight-six (Triumph-powered) car which was more practical and successful than any of the others.

Although one car led to another, it was not always an easy or logical progression, so each model will be analyzed and described on its own, and be related to others in the same general family.

The Griffith 200 and 400 models

For British readers of this book, the first point to clarify is that the vast majority of all the Griffiths were exported to the United States. It would be very surprising if more than a handful of V8-engined TVRs are still in the UK.

Compared with the chassis and suspension of the Grantura Mark III, on which the US-built prototype had been developed, there are several significant differences. The basic chassis-frame was unchanged – neither tube diameters nor the wall thickness of the tubes were increased – but suspension pick-up points were strengthened. In place of the

Warren Allport, of *Autocar*, peering over the scuttle of the 1965 Griffith 400 which the magazine tried briefly at the beginning of that year. Apart from the large and prominent bulge in the bonnet moulding, there was virtually no other indication that this car had a V8 Ford engine. (*Autocar*)

Grantura Mark III's secondary damper to each side of the rear suspension, a second combined coil spring/damper unit was attached. Spring and damper settings were revised, not only because of this, but to take account of the extra weight of the 4.7-litre V8 engine mounted up front. Surprisingly enough, the original 200-Series Griffiths retained the BMC B-Series final-drive, and the driveshafts to match, which seemed to cope remarkably well in spite of the huge overstressing which they received.

The visual changes were obvious. At the front there was a sizeable bonnet bulge, made necessary to clear the air-cleaner of the large American engine and at the rear there were twin tail-pipes to the new exhaust system, which obviously had been necessary. Fatter-section (185) radial-ply tyres were specified, and the 72-spoke wire wheels had a 5in rim width.

There were two degrees of engine tune – rated at 195bhp (gross) and 271bhp (gross). The 195bhp engine gave birth to the car's '200' title, it is thought, and this was a real mass-production unit to be found in several other American Fords of the day and in the AC Cobra (or AC 289 as it would also eventually be named). The 271bhp engine was a truly sporting unit, available at extra cost on several such cars, and one also destined to be the most powerful unit offered on the Ford Mustang of 1964.

Subsequent experience suggests that the engine outputs equated to something like 150bhp and 220bhp (nett) respectively, which doesn't sound as exceptional, but the fact remains that the performance of a Griffith was quite phenomenal by any standards. Certainly, in straight-line performance, if not in ultimate road behaviour, the car was good enough to embarrass severely an E-type Jaguar and many V12-engined Ferraris!

When Jack Griffith did the deal with Grantura Engineering, not only was he in a great hurry to get some cars delivered and on sale, but he insisted that they should be supplied to him without engines and gearboxes (the gearboxes, by the way, were the same type as those used in the Cobra), so that he could have them fitted in his own workshops on the East Coast of the United States.

It is only fair to say that Grantura Engineering had neither the capital, time, nor resources to cope efficiently with the technical and commercial demands of Jack Griffith. The result was that production got under way slowly, and the original cars were by no means properly developed. The main problem was that the engines were often found to overheat – which is hardly surprising when the small size of the bonnet air intake (the same as that for the Grantura Mark III) is considered.

No-one can now say how many of the original Griffith 200s were built, but what is certain is that the machine which followed – the Griffith 400 – was a better all-round car. Firstly,

Very few Griffith 200s – those with the Grantura Mark III tail treatment – were ever built. This is one of them, without its engine and transmission, about to leave for the construction to be completed in the United States. The company title on the wall was already obsolete, for the picture was taken in 1963 and TVR Cars Ltd had descended into bankruptcy in 1962. (*John Bailie*)

Typical of a crowded engine bay for any of the Ford V8-engined TVRs of 1967. The tubular chassis-frame needed to be modified to accept the great width of the 4.7-litre engine, but the only major body change required was to incorporate a hump in the bonnet moulding to give clearance to the air-cleaner. (*John Bailie*)

Martin Lilley's own short-wheelbase Tuscan V8 under test by *Motor* magazine in the spring of 1967. 500ML, which was nicely restored in the hands of a TVR Car Club member, had a 271bhp engine and was shatteringly fast; its top speed was estimated at more than 155mph by *Motor*, and it was able to sprint to a quarter-mile from rest in a mere 14.1 seconds. Not even an M-Series Turbo could match that. In all respects except for the bonnet bulge, a Tuscan V8 looked like a Grantura 1800S, with which it was concurrent. (*Motor*)

and most obviously, it was the original TVR to be treated to a cut-off Manx tail, complete with circular Ford Cortina Mark I tail-lamp/indicator units, and the first of these cars was built in the early spring of 1964. In addition, the Griffith 400s were also fitted with thermostatically controlled Kenlowe cooling fans, which – together with a much larger-capacity water radiator – helped to cure the overheating problems.

The Griffith 400s were exhibited at the New York auto show of April 1964 (incidentally, this was the same show at which the 4.2-litre Ford V8-engined Sunbeam Tiger also made its debut). Soon the engineless cars were being turned out by the Blackpool factory at five a week, a rate which doubled by the end of 1964 (and sometimes approached 15 a week); this was the highest rate of TVR building ever achieved until the Nineties.

The 400, however, was not a properly developed car, especially the version which was fitted with the optional 271bhp engine. The Griffiths often showed off the worst traits of cars which were not only designed and built in a great hurry, but were also assembled without due attention to quality of construction. Complaints piled up – not least from Jack Griffith to the factory – and the car's reputation began to suffer.

Until the beginning of 1965, each and every Griffith had officially been sent to North America. However, at the 1965 Racing Car Show, which was held at Olympia in January, the Griffith was made available in the UK, priced at £1,342 (basic), or £1,620 (total) for the 195bhp model, £1,488 and £1,797 respectively, for the 271bhp model.

The facia of the short-wheelbase Tuscan V8, crowded with instruments and switchgear, was entirely different in detail from that of the Grantura 1800S, though both cars used the same type of Triumph Spitfire/GT6 steering column. (*Motor*)

This heavily retouched artist's impression is of the Tuscan V6 – basically a Tuscan V8 with a 3-litre British Ford engine. The sidelamps drawn in for this study were never used, but the engine bay vents and the depressed nostril air intakes were. The long bonnet bulge was needed to clear the V6's air-cleaner. Cast-alloy wheels of this pattern were introduced specially for the Tuscan V6, and the style was later adopted for the Vixens and 2500s. (*John Bailie*)

In the meantime, the Anglo-US business relationship concerning the Griffith was shattered by a prolonged dock strike in the USA, which paralyzed British exports to that country and ruined the finances of both Grantura Engineering and the Griffith assembly business. It was the beginning of the end for the Griffith and a major factor in the eventual demise of Grantura Engineering.

Although Grantura collapsed during 1965 (and the production of the Griffith failed with this collapse), it was not quite the end of the mighty V8-engined car. Once the Lilley family had taken over the remains of the TVR business and had got things going again, the existing models were put back into production. As related in Chapter 3, the major activity in 1966 involved the Grantura 1800S models, but the new concern, TVR Engineering, also managed to turn out 10 Griffiths. Therefore, although it can only be said that *approximately* 300 Griffiths were produced by Grantura Engineering between 1963 and 1965, it can be asserted that *exactly* 10 were assembled by the new concern.

A study of the Chassis Books show that the first of these cars was built in April 1966 and the last in January 1967. Four had right-hand drive, while six had left-hand drive. Incidentally, despite the fact that the car had officially been a Griffith 400, since the Manx-tail style was adopted, the Chassis Numbers of these final cars were prefixed with the number 200...

Although it is impossible to be definite about this, it is estimated that no more than 20 Griffith models were delivered in the UK – all the rest originally went to the USA, and this is where most of the survivors undoubtedly exist today.

The Tuscan V8

Mechanically, the car which took over from the Griffith *was* a Griffith, though it was renamed in an unsuccessful attempt to kill off the Griffith's bad reputation. Indeed, the new car, called a Tuscan V8 by Martin Lilley, was so similar to the Griffith that it took up the same sequence of Chassis Numbers: the last of the Lilley-management Griffiths was 200-010 and was built in January 1967, while the very first Tuscan used 200-011 and followed on immediately.

Like the Grantura 1800S, the Tuscan V8 was treated to

higher quality fittings and had a highly polished wooden facia panel. However, this was quite different in detail from that fitted to the 1800S. If the car was fitted with the 'soft' 195bhp engine it was known as a Tuscan V8, but if the optional 271bhp engine was specified it became a Tuscan V8 SE (Special Equipment).

Even though TVR had appointed a new USA distributor (Gerry Sagerman, who had indirectly been responsible for the birth of the Griffith in 1963), the sad story of the Griffith had destroyed the reputation of this car in that continent. Despite the name change from Griffith to TVR Tuscan V8, it was all too obvious that the two cars were basically the same. The Tuscan V8 had received some development, notably to the cooling system, but it was still by no means cheap.

In the UK, in 1967, the 195bhp Tuscan V8 was priced at £1,967 – which was exactly the price asked by Jaguar for an open E-type – while the 271bhp Tuscan V8 SE cost £2,364 – which was £80 *more* than that asked for the svelte 2+2 E-type Coupe. That was the nature of TVR Engineering's problem that year, and it showed in the sales achieved. In six months a mere 28 Tuscan V8s were built, of which just four had right-hand drive.

In the meantime, however, Martin Lilley had been developing the basic concept, and the Tuscan V8 long-wheelbase model was the result.

The Tuscan V8 SE LWB

Even as the Tuscan V8 was being launched at the beginning of 1967, work was going ahead to produce its successor. This was significant because it was the first attempt by the new management to update and improve the Grantura chassis and body, which had originally been joined together in 1962. Not only did the new car have a lengthened chassis, but it was also significantly restyled.

The multi-tubular chassis had its wheelbase lengthened from 7ft 1.5in to 7ft 6in and this figure was not changed again until 1980, when the long-running M-Series models were finally withdrawn. Due to the simple nature of the layout, this stretch was fairly easily arranged, and its purpose was to allow the doors to be lengthened and for more space to be built

Retouched, but still representative of the Tuscan V6 model built from 1969 to 1971, which used the Manx-tail body first seen in 1964 and discontinued in 1972. The wrap-around Ford Cortina Mark II tail-lamp clusters were a feature, and the twin tailpipes plus Tuscan V6 badging made this particular TVR quite unmistakable for another. Like all TVRs built from the 1968 model-year onwards, the Tuscan V6 had a 7ft 6in wheelbase. (*John Bailie*)

into the passenger compartment. The suspension layout and almost every mechanical fitting connected to the chassis was unchanged. The extra 4.5 inches went into lengthening the longitudinal chassis tubes and, of course, a longer propeller-shaft and exhaust system were required. The lengthened chassis, incidentally, had originally been conceived for the Trident project, which is described in Chapter 7.

To match the lengthened doors, the long-wheelbase Tuscan V8 SE was also given the latest Ford Cortina Mark II tail-lamp/indicator assemblies (which replaced what are affectionately called the 'Ban the Bomb' style of circular assemblies from the Cortina Mark I) and a different bonnet moulding, which included a longer and more shapely central bulge (to clear the engine air-cleaner) and different air vents.

Appropriately, the Chassis Number sequence was changed for the new car, which took on the LWB prefix. They were no more successful than the original short-wheelbase Tuscan V8s, however, for only 24 machines (one a rolling chassis, exhibited at the Earls Court motor show) were built in about a year. Half of them had right-hand drive.

However, this car gave rise to another, even more significant model, which was also a Tuscan V8 SE, but had a quite different bodyshell. What shall be called here the 'wide-body' Tuscan V8 LWB was announced at the 1968 New York auto show and carried the chassis prefix of MAL (Martin Lilley's initials – for it was Martin who was mainly responsible for the restyle). Between April 1968, when the first car was built, and August 1970, only 21 of these cars were assembled, all but two of which had left-hand drive.

It was the first time that the TVR coupe body style had been comprehensively reworked. Based on the lengthened (7ft 6in wheelbase) chassis and retaining the same glass as the existing models, the new body was four inches wider below the waistline and had completely new and much smoother lines around the nose and the tail. At the rear, the Mark II Cortina tail-light clusters were neatly moulded into the corners, and at the front there were new sidelamps. New bumpers completed the picture of what was a very smart new car.

Although it required new moulds and was more comprehensively equipped than any previous TVR, Martin Lilley made no attempt to fit this bodyshell to the lower-powered TVRs and, as a consequence, it was destined to become a loss-making project.

It was, however, a real pointer to the future, and certainly formed the theme on which the definitive M-Series body style was based. Note, though, that the last of the 'wide-body' Tuscans was built in the summer of 1970, while the first-ever M-Series car was not completed until the autumn of 1971, and series production did not begin until April 1972.

The Tuscan V6

Once the TVR marque became re-established, it was clear to Martin Lilley that there was a huge marketing gap between the 1.6-litre four-cylinder-engined Vixen and the 4.7-litre V8-engined Tuscans. However, it was not until 1969 that the new company, TVR Engineering, was in a sufficiently healthy condition to consider another additional model. The result was the Tuscan V6, first put into production in mid-1969 and price listed (at £1,492 basic, £1,930 total, in the UK) from May 1969.

The fact that its Chassis Number sequences were prefixed by the letters LVX (as for the Vixen Series 2 and Series 3 models and effectively meaning Long Vixen) accurately denotes the Tuscan V6 as being an extension of the Vixen theme, rather than a 'tamed' version of the slow-selling Tuscan V8s. The name, in a way, was more of a misnomer than might at first be thought.

By 1969, TVR Engineering had satisfactory commercial links with the Ford Motor Company (they had been trading with them since the late Fifties) and there was little difficulty in fixing up supplies of Ford Zodiac V6 engines and gearboxes for their new car. (Note that, strictly speaking, the Tuscan did not use *Capri* 3-litre units at first, as this model was not officially announced until October 1969, several months after the Tuscan V6 went on sale.) These engines, incidentally, were almost identical with those supplied by Ford to other companies like Reliant, for the Scimitar GTE, and Gilbern, for the Genie/Invader models, not forgetting TVR's main rival, Marcos.

The rolling chassis was effectively that of the Vixen Series 2, for only the one combined coil spring/damper was used (along with a secondary damper), though the servo-assisted brakes of the Tuscan V8 were standardized and the cast-alloy wheels had a 5.5in rim width. Compared with the Vixen, which had 88hp (nett), the Tuscan V6 had 136bhp (nett) or 128bhp (DIN) as Ford liked to have it quoted, which represented a more-than-50 per cent increase in maximum power and torque. The gearing of the Tuscan V6 was considerably higher than that of the Vixen (its final-drive ratio was 3.31:1 at first, 3.54:1 later on, compared with 3.89:1 for the Vixen), but it was still a car which accelerated much more smartly, and it had a maximum speed of up to 125mph.

It is also indicative of how much of a hybrid the Tuscan V6 really was that it retained the massive Salisbury 4HU differential and final-drive unit (as fitted to the V8-engined cars), rather than used the smaller BMC or Triumph differentials favoured by the four-cylinder cars. There is a complete list of differential fittings – units against model – at the end of Appendix B, which covers technical specifications.

The Tuscan V6 could be recognized by its badging on the Manx tail (the widened body of the final Tuscan V8s was not used), its twin exhaust tailpipes and its distinctive exhaust note. It was an acceptable compromise in many ways, but it was not an outstandingly successful car. At a time when Vixen sales were reaching towards six cars a week, those of the Tuscan V6 never achieved even two cars a week. It was in production from mid-1969 until the end of 1971 (ie, just before the last of the 1962–72 chassis was assembled), during which time a total of 101 cars were built and sold.

The main problem was that its engine was not 'de-toxed' for sale in North America, where the majority of TVR's exports were concentrated. TVR Cars of America Ltd, founded by Gerry Sagerman in 1966, were now making headway with the rehabilitation of the marque, and they needed a model with an engine which could meet the ever-tightening demands of the North American legislators. The Ford V6 engine was not being exported to the USA (though the new Pinto engine was scheduled to go, and the existing Vixen's 'Kent' unit was also on sale there in some Ford models), and Ford had no plans to have the engine redeveloped to qualify. TVR Engineering, for their part, simply could not afford to do the job themselves.

Therefore, when the Tuscan V6 had only been on sale for about a year, work began at Blackpool on yet another derivative – one which would, above all, be aimed straight at North America; V6 power was ditched and straight-six power was adopted. The result was the TVR 2500, the first-ever TVR model to have a straight-six-cylinder engine.

The British Ford V6 3-litre engine slips neatly into the frame and engine bay which was also large enough to accept the rather more bulky American Ford 4.7-litre V8. TVR cognoscenti will notice yet another derivative of the 'where shall we put the header tank' theme. Note the direct-acting vacuum servo for the brakes (ahead of the toeboard) and the battery in its habitual position to the left side of the underbonnet space. The 3-litre engine became the mainstay of TVR cars of the Seventies; this was its first use, in 1969. (*John Bailie*)

A very exclusive (and expensive-to-build) TVR of 1968–70 was what is now known as the wide-body Tuscan V8. This car used the longer-wheelbase (7ft 6in) chassis and the powertrain of the Tuscan V8s, but was treated to an entirely different and smoother bodyshell, which Martin Lilley helped to style. Its nose was longer, its tail was smoother and it was four inches wider below the waistline. (*Autocar*)

In 1970 TVR moved its factory from the Hoo Hill building to Bristol Avenue, but there was still not much room to move. There are few clues, but I date this shot as 1971, at Bristol Avenue, for those are 2500 chassis being assembled – and the rear-end body style is of the Tuscan V6/2500/Vixen S3 variety. (*TVR*)

One of the early 2500M models of 1972, showing off the distinctive lines of the M-Series body, clearly but not directly developed from those of the wide-body Tuscan V8s of 1968–70. Recognition points include the engine size marking, the early type of cast-alloy wheels, the Zodiac Mark IV type of engine bay cooling hole grille and the chrome bumpers. On these early cars, too, the wrap-around Cortina Mark II rear-lamp clusters were retained. (*John Bailie*)

The TVR 2500

The key to the success of the new car was its engine. TVR cast around for a unit already selling in the United States, one which was also scheduled for persistent development with a view to it continuing to meet the regulations in future years. They found just such a unit at Triumph – the de-toxed 2,498cc unit, an overhead-valve straight-six which had first been used in the TR250 of 1967 and was the engine currently powering the TR6 Carb model built exclusively for North America. It was neither as powerful nor as torquey as the Ford 3-litre V6, but it was fully certified and likely to remain so. As supplied by British Leyland (Triumph had been a part of that combine since it was founded in 1968) the engine produced 106bhp (nett) at 4,900rpm and came complete with a sturdy all-synchromesh gearbox and optional overdrive. Also in the package, as developed by TVR, was the chassis-mounted differential and final-drive unit from the same Triumph TR6 model, one which became of supreme importance to TVR throughout the Seventies.

It all came together in the nick of time, for the Tuscan V6 was quite unsuitable for export (only seven cars were sold overseas) and the Ford Cortina/Capri 1,599cc unit was just about to run into several development problems regarding the USA de-toxing requirements.

The TVR 2500 used the Tuscan V6/Vixen S3 rolling chassis, with Tuscan V6-width 5.5in wheel rims, and had a smaller fuel tank – to take account of yet more evaporative-loss regulations – than the Vixen or Tuscan models. Although

One of the only two right-hand-drive wide-body Tuscan V8s in existence is this extensively restored example, owned by Robert Nosowicz, which used to be Martin Lilley's own car. This photograph (taken in 1980) shows a much-modified 298cu in (4,727cc) V8 engine, special wheels and tyres and a detachable Perspex roof panel. The car has been used in sprints and other competitions, which explains the number on the door. (*Robert Nosowicz*)

it was shown at the Earls Court motor show in October 1970, it was not quite ready to be put on sale. The first production model, Chassis Number 1745/6T (6T = six-cylinder Triumph) followed in November, and the full flood of production began at the turn of the year. It was not until April, however, that British magazines were issued with an interesting overhead picture showing the Triumph-engined TVR 2500 rolling chassis, alongside and in front of a complete 2500 car. The UK price was £1,475 (in kit form, without Purchase Tax) and £1,927 (complete), though these prices began to escalate before long, as Britain's Seventies' inflation had now begun to worsen alarmingly.

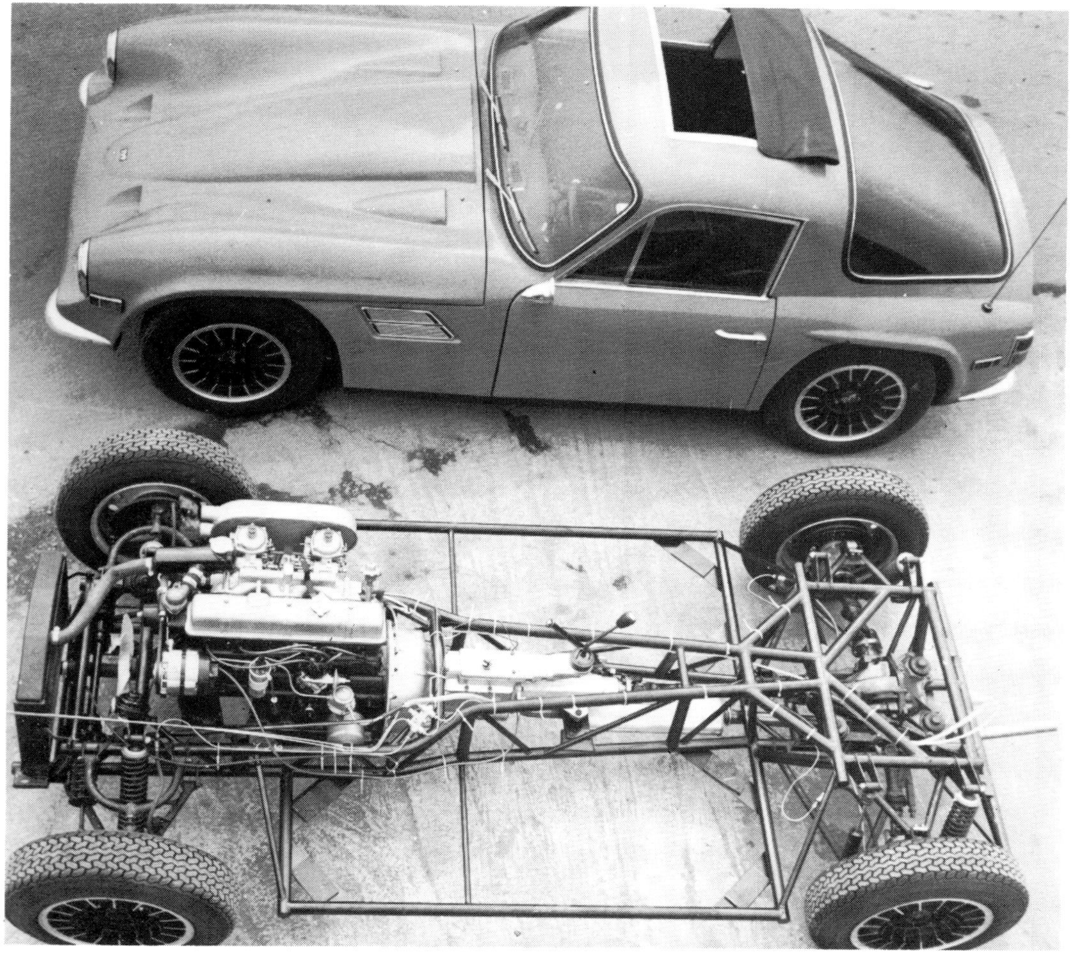

A splendid picture originally issued by TVR in 1971, which shows off not only the relationship between the late-model Vixen-type coil-sprung chassis and the smoothed-out bodyshell used between 1969 and 1972, but also the detail of the chassis, and the way in which the de-toxed six-cylinder 2,498cc Triumph TR6 engine and gearbox fits into the frame. The contours of the longitudinal tubes by this time were fixed by the need to clear the bulky Ford V6 and V8 engines. In this case, the final-drive assembly is from a Triumph TR6. The sunroof on the car was an optional extra, and the rear-facing bonnet air outlets were to keep the 'federal-ized' engine adequately cool. (*TVR*)

Overhead view of Robert Nosowicz's ex-Martin Lilley wide-body Tuscan V8, showing off the large aperture in the glassfibre roof, which is normally filled by a Perspex panel. The engine has been heavily modified and features four downdraught twin-choke Weber carburettors. (*Robert Nosowicz*)

Surprisingly, TVR and Martin Lilley were never tempted to sell the 2500 in non-US market guise with the much more powerful fuel-injected TR5 engine (which produced peak power of 150bhp at 5,500rpm), and the result was that the 2500, though a lively enough car, was not markedly quicker than the Vixen S3, which continued to be very popular.

The 2500 was a great success by any previous and current TVR standards. Although only eight cars were started in 1970, no fewer than 237 were assembled in 1971, compared with 97 Vixens, and production was still going ahead at full-blast when the last of the Grantura-type chassis was assembled in the spring of 1972. Not only were a further 44 original-type 2500s built in the first few months of 1972, but a further 95 cars of the hybrid type were built in 1972 with the new M-Series chassis and the old-style bodyshell, and another of these 'hybrids' was built in 1973.

It follows from these remarks that there was a considerable overlap during 1972 (and, to a lesser extent, in 1973) between the 2500, which had the old type of chassis-frame, the hybrid 2500 which linked the old-style body with the M-Series frame, and the 2500M, which used the familiar drivetrain in the M-Series frame topped by the M-Series body style. All very confusing.

Of the 2500s, therefore, 289 were completely old-style cars and 96 were M-Series chassis/old-style body cars – making a total production of 385. Of these, at least two-thirds were exported to North America.

The last of the completely old-style 2500s was Chassis Number 2237/T, which was not quite the last of all the old-style frames, and the last of the hybrids was 2703/T, commissioned in May 1973.

In detailing the development of the more powerful TVRs of the Sixties and early Seventies, the major business change which occurred during the period has been ignored. Early in 1970, negotiations began for a new factory to be acquired at Bristol Avenue, Bispham, Blackpool, and at the end of that year the move was completed. It allowed TVR's plans for the Seventies to be pushed ahead – somehow, at the old Hoo Hill works, the future had always seemed to be such an ephemeral vision – and these included the design of another new chassis *and* body style. The result was the M-Series cars, where M stood for Martin. If the Fifties had been all about learning and the Sixties all about survival, the Seventies were to see TVR come to maturity.

CHAPTER 5

1600M, 2500M and 3000M

M for Martin – the Seventies' TVR

As far as motoring journalists were concerned, a visit to the TVR stand at the 1971 Earls Court motor show was a memorable experience. Not only were there two new TVR models to be seen, but the press day celebrations were also enlivened by the presence of two nude models cavorting around the cars. It brought TVR more publicity, of one kind or another, than they had ever expected, though it is doubtful if even one per cent of all the pictures which were taken ever found their way into print.

The two new cars on show were the Spitfire-engined 1300, which had been in production since August 1971, and the 2500M, which was just a prototype rushed to the show immediately after it had been built at Blackpool. That car, incidentally, carried Chassis Number 2090T; it is not known if it still survives. There was also the SM, or Zante prototype, but that story is covered in Chapter 7.

The 2500M's launch, in fact, was curiously muted, especially as it proved to be the cornerstone of TVR's prosperity in the Seventies. I well remember seeing the car, noting that it had different front and rear styling and wondering what all the fuss was about. At the time, little was made of the car's most significant feature – its new chassis – perhaps because this wasn't even fitted to the show car, but was added later.

The design of an all-new chassis for the cars of the Seventies signalled to TVR watchers that the Lilley family now had long-term confidence in the marque which they had bought in 1965. They were ready to make a large investment in TVR Engineering, not only to provide a new image, but to help eradicate some of the shortcomings of the past.

By 1971, the existing TVR chassis, which had originally been used in the Grantura Mark III of 1962, had been in production for more than nine years. In that time, its wheelbase had been lengthened once, the profile of the tubes around the engine bay had been modified to take account of the various engines used over the years, and it was still a frame which was more or less hand-assembled by skilled craftsmen. Not only this, but it had always had something of a reputation for steering kick-back and other development maladies.

In the meantime, Mike Bigland, an engineer and TVR dealer from the Midlands, had built up a special Tuscan V8 for a customer, incorporating many chassis modifications. After a visit to see Martin Lilley at Blackpool, Bigland got together with Martin to evolve a new chassis design incorporating some of those ideas and to form the basis for the new range of models. This soon became known as the 'M-Series' design, and as already mentioned, M stood for Martin.

Although the prototype frame was completed in the autumn of 1971, the factory was not yet ready to produce it in quantity. Existing Vixen-type frames continued to be built in large quantities for several more months after the 2500M prototype was shown at Earls Court, and the first 'off-tools' example was not built until April 1972. At that time, however, there was a simple and straightforward functional – and statistical – changeover. The last of the old-style frames was Chassis Number LVX/2239/4, and it formed the basis of a Vixen Series 3, while the first of the M-Series frames was Chassis Number 2240TM, and it became the original 2500M

The engine bay of the 2500M, the first TVR built with a straight-six-cylinder engine. This was the de-toxed Triumph TR6 unit, which produced about 106bhp. On all M-Series cars the spare wheel was carried in the nose, ahead of whatever engine was fitted, and above the cooling radiator. The trunking for the heater installation was now to the nearside of the engine bay, and the bay on the 2500M was cluttered with pipework seemingly inseparable from the demands of the exhaust-emission regulations, which intensified as the Seventies progressed. (*TVR*)

The original-style 'chrome-bumper' 3000M, showing off its characteristic lines, but running here on non-standard Wolfrace light-alloy road wheels. (*John Bailie*)

production car. There was no overlap and no confusion.

As with all previous TVR production cars, the new M-Series chassis-frame was a multi-tubular design, where four longitudinal members linked front suspension to final-drive and rear suspension mountings and formed a sturdy box-section backbone. This core was linked to perimeter members encircling the passenger compartment floor, and there were suitable reinforcements around the scuttle, in the nose and at the rear. So that the old-style bodies could be used for an interim period, and because Martin Lilley could see no reason to enlarge a two-seater TVR any further at the time, the M-Series chassis' wheelbase remained at 7ft 6in (the figure established in 1967), though front and rear track dimensions were marginally increased.

The big difference was that the traditional 16-gauge 1.5in

Interiors changed in detail from time to time, but this was typical of the 1600M/2500M/3000M of the mid-Seventies. It shows clearly the very high transmission tunnel, the shape of which was determined by the position of the inertia reels for the safety belts, and the reason why there could be no question of '+2' seating in these cars. (*John Bailie*)

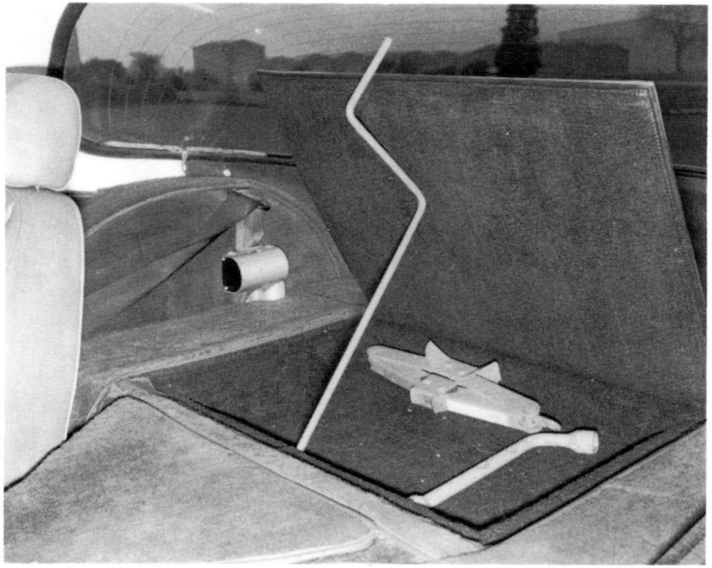

The stowage of the wheelbrace and jack in the M-Series cars was under a lifting panel in the loading area, behind the seats. This particular car was a 1976 1600M, on test. (*Autocar*)

diameter tubing was only retained in some places. In others, tubes of a thicker (14-gauge) wall section were chosen and in some sections of the frame there were square-section tubes; this was the first time that square-section tubing had been used by TVR for other than local reinforcement, and it made the structure simpler to build and somewhat easier for the bodyshells to be attached.

The front suspension was much as before, though the wishbone layout was made more sturdy by the use of the longer and more robust Triumph TR6 uprights. The steering rack was by Alford and Alder (as used on Triumph models) and the collapsible steering column was also a BL part. The independent rear suspension was similar to the superseded Vixen-type layout, but in the M-Series cars the combined coil spring/damper unit was mounted ahead of the line of the driveshafts.

The whole frame looked more sturdy than ever before, for

the four-tube backbone section was cross-braced with round and square-section tubing. It was, indeed, a frame which could easily take care of the burgeoning crash test legislation which, by this time, had spread from the USA to Europe and other territories.

A truly thoughtful feature of the new layout, made possible by the new body style, was that the spare wheel was stowed in the front of the engine bay, ahead of the engine, and above the wide, shallow water radiator. This meant that the luggage space behind the seats was now completely unencumbered by a wheel and its cover, a disadvantage which had been present in all TVRs built since 1964 with the Manx-tail style. The road wheels were in cast aluminium alloy, with 15in rims and a 5.5in rim width, initially to the same style as those already in use on 1971–72 non-M-Series cars. As before, the braking system was by Girling, using 10.9in front discs and 9in rear drums, with vacuum-servo assistance.

As far as the public was concerned, the most obvious

A detail of the front of a 3000M, showing the Wolfrace wheels, which were only standard on the 10 special 1976 'Martin' Anniversary cars, the inner wheelarch, which kept the engine bay clean, and the rather limited opening angle of the long-nosed M-Series bonnet moulding. This is one of the later 'black-bumper' cars. (*TVR*)

Impressive view of the 3000M rolling chassis, polished and prepared for exhibition at the 1973 Earls Court motor show. By this time, the latest pattern of light-alloy road wheel, with 14in rim diameter, had been adopted. This view emphasizes the tightly nipped-in waist of the M-Series backbone chassis-frame and shows another recognition point – the mounting of the rear suspension's coil-spring/damper units ahead of the driveshafts. The car in the background is a 2500M, and it is just possible to see that the Triumph TR6 type of tail-lamp clusters had been adopted by this time. (*TVR*)

Overhead picture of the rolling chassis and engine installation of the 3000M of 1973, showing the very neat layout of the coil-sprung front suspension, the radiator position, the mounting of the spare wheel (which was, incidentally, something of a safety feature) and the very sturdy arrangement of square-section and round-section chassis tubes in this part of the frame. The tyre section was 185HR-14in. (*TVR*)

feature of the M-Series car was the new body style. Although the new cars were still two-seaters, with no attempt even at providing 'plus-two' accommodation, and retained the same windscreen, doors and rear-window profile, they were new in most other respects and looked to the 1968–70 wide-body Tuscan V8s for their inspiration. At the front, the bonnet was longer and sleeker, partly because Martin Lilley liked it that way, and partly so as to accommodate the front-mounted spare wheel (which was a useful aid to providing barrier-crash resistance), with a more rectangular air-intake than hitherto to match the radiator profile. At the rear, the lines were smoothed-off very like those of the wide-body Tuscan V8s, and incorporated the Cortina Mark II tail-lamp clusters, which were now obsolete as far as Ford were concerned, for their model had been changed in the autumn of 1970.

Inside was a plush new interior, made instantly recognizable by the wide, high and flat propshaft tunnel created by the

The well-filled 3000M engine bay, complete with its Capri/Granada 2,994cc British Ford V6 engine, which could produce up to 138bhp, should be compared with the Tuscan V6 installation shown elsewhere in this book. Among the many interesting details are the cable-operated clutch, the direct-action vacuum brake booster, the separate overflow-type tank for the cooling system and the multiplicity of pipes, hoses and wires necessary in every modern car. (*John Bailie*)

layout of the tubular backbone frame. The speedometer and the rev-counter were directly ahead of the driver, while the auxiliary instruments were centrally mounted.

On the original cars, the bonnet was graced by a pair of air-intakes – one on each side of a central bonnet bulge which was smoothly blended into the nose, just behind the TVR badge – and the side of the bonnet still incorporated extractor vents capped by the decorative Aeroflow grilles from the Ford Mark IV Zephyr/Zodiac models.

The new M-Series body style was the same width and height as its predecessor – for these dimensions were effectively ruled by the use of the screen, rear window and doors from the old model – but in its finalized form it was nine inches longer than before, at 12ft 10in instead of 12ft 1in, with most of the increase going into the front overhang, though there was a useful, if marginal improvement in the stowage space behind the seats. One feature which was *not* improved, however, was access to that stowage space. M-Series cars, like all previous TVRs, were not equipped with an opening bootlid or rear window, a failing which began to have more impact on road testers' opinions as the Seventies progressed.

Although it was only the 2500M prototype which was displayed – and talked about – at Earls Court in 1971, TVR's plans for the use of the M-Series chassis and body style were far-reaching. An integrated range of three cars – the 1600M, the 2500M and the 3000M – was planned, and all three cars were to be launched to the British press in June 1972.

Talking about it, however, was easier than achieving it. At the time of the British launch of the three cars, only the 2500M was truly in series production, for the first 1600M (2288FM) was only then being built and the first 3000M (2410FM) was not actually started until September 1972. At the same time, too, the scene at Blackpool was confused by a handful of M-chassised 1300s, a short run of Series 4 Vixens (M-chassis, Vixen S3 body) and a considerable quantity of 2500s (now with M-Series chassis, but Vixen-style body). It

A special high-luxury version of the 3000M, known as the 3000ML, was introduced at the end of 1973 and ran for the following year. External identification includes a black vinyl roof, sliding roof and matt black valances and side sills, while inside there was a walnut dashboard and more luxurious upholstery and trim. (*TVR*)

This magnificent cutaway drawing of the 3000M was prepared by John Bailie, a TVR owner who was also the proprietor of Image Publicity, who had close business connections with TVR Engineering Ltd. This shows the 3000M in its definitive state, as the most successful of the Seventies' TVRs sold in the UK. Note the cylindrical fuel tank, mounted low down and at the rear of the car. (*John Bailie*)

was TVR's most prolific period, and one which is now remembered with some incredulity.

Once the change had been completed *in toto* – and this happy state was not achieved until the spring of 1973 – everything appeared much more logical, with three models being based on the same basic chassis and suspensions and using the same two-seater coupe body style, identical in all respects except for badging. The 1600M had taken the place of the Vixen Series 3 and Series 4, the 2500M had replaced the 2500, and the 3000M had effectively replaced the Tuscan V6, which had died before the end of 1971.

At this point, it should be recorded that Britain's negotiations to enter the European Economic Community (or 'Common Market', as it is more usually known) had been successfully brought to a conclusion in July 1971, and that as a consequence the British Purchase Tax system was scrapped early in 1973, soon after the country actually joined the EEC, to be replaced by the Value Added Tax system already used in other Common Market countries.

The significance of this was that from March 1973 there was no longer any advantage in buying a car in kit form, as VAT was charged even when the car was sold in an unassembled state. Even though for some time Martin Lilley's policy had been to discourage the sale of TVRs in kit form (he could be more sure of the completed cars' quality and reliability), some TVR kit cars were still being sold in

The 1600M, something of an orphan at TVR, introduced in 1972, dropped in 1973, reintroduced in 1975 and dropped finally in 1977. This particular car, built in 1976, is visually typical of other M-Series cars being produced at Blackpool at that time. (*TVR*)

1972 and 1973. The effect of the arrival of VAT was to kill off that demand instantly, with an upset in the demand for certain models.

The two other major factors influencing TVR's development in the early and mid-Seventies were the continuing (and, indeed worsening) series of strikes at Triumph, which affected the supply of engines for the 2500M, and the disastrous fire which swept through the final assembly building on the evening of Friday, January 3, 1975.

Each of the three successful M-Series coupes will now be studied and assessed in the order of engine size.

The 1600M

This was a model which had a rather stop-start history at TVR. It was introduced along with the other M-Series models in June 1972 but was withdrawn less than a year later. Then, following the factory fire, and in response to a change in market conditions, it was reintroduced in May 1975 and was in production for a further two years. Effectively, the 1600M was a direct replacement for the very popular Vixen series, though it must be pointed out that the 'interim' Vixen Series 4 was on offer throughout the period when the 1600M was first in production. The only significant difference between the two types was that the 1600M was fitted with the M-Series body style and the M-Series chassis, while the Vixen Series 4 used the old-style Manx-tail bodyshell atop its M-Series chassis.

A price comparison is interesting. When the 1600M was introduced in June 1972, its basic price in the UK was £1,560, whereas that of the Vixen Series 4 was £1,425. At the same time, incidentally, the placarded price of a 3000M was £1,795 (even though this model was not yet in production). Nine months later, when the last of the Vixens *and* the first run of 1600Ms were being built, the same basic prices were: Vixen Series 4 £1,496, 1600M £1,638 and 3000M £1,884.

The 1600M used the same rolling chassis, bodyshell and fittings as the other 1972–73 M-Series cars, the only obvious difference being that the 1,599cc Ford 'Kent' engine, in 86bhp (DIN) Capri GT tune, along with the excellent four-speed all-synchromesh gearbox, was fitted. So much was

Even with the relatively simple four-cylinder 1,599cc Ford Capri engine fitted (as the 1600M), the TVR M-Series engine bay is well-filled. There is one downdraught compound-choke Weber carburettor, almost hidden by the big air-cleaner, and the exhaust system is on the left side of the engine. (*Autocar*)

M-Series production of the mid-Seventies with the Bristol Avenue scene completely dominated by Triumph six-cylinder 2500M engines in the foreground and lines of Ford V6 3000M engines behind them. This represented about three weeks' supplies for the production lines, which rarely built more than 10 cars a week. (*John Bailie*)

obvious to the casual glance, except for the use of the Triumph GT6/Vitesse type of differential and final-drive casing. The weight of the car was 1,970lb unladen, which compared with 2,240lb for the 2500M and 3000M models.

At first, to TVR's consternation, the 1600M was not a success. In 1972, a total of 57 1600Ms were ordered into production, but in the first three months of 1973 they were followed by a mere 11 cars. It was a time, of course, when there was great demand for the 2500Ms and the 3000Ms, but there must surely have been other reasons.

Stewart Halstead, when TVR's Sales Director, recalled: "When the M-Series models were introduced in 1972 we were still making kit-cars for British customers. Then, when VAT came in, early in 1973, the home market reduced drastically. All the 2500Ms – or nearly all – were going to the States, and we decided that the British-market TVR should be the 3000M."

The change from Purchase Tax to VAT had a more profound effect on specialist manufacturers like TVR than many of them have ever admitted. In 1972, TVR had sold 220 cars in the UK, but in 1973 this plunged to 101. One effect of this was that more capacity was available for export-market models (particularly the 2500M to North America), where sales increased from 168 to 287. It was a trend which Martin Lilley encouraged in 1974 for, following the 1973 Suez crisis, home sales of TVRs dropped still further to a mere 36 cars, while export sales boomed ahead to 385. It may sound paradoxical, but in 1974, the 'post-Suez' year when the world's largest manufacturers all began to suffer the effects of vast increases in fuel costs, TVR sold more cars than ever before.

The first and last 1972–73 1600Ms carried Chassis Numbers 2288FM and 2623FM, and there was then a 26-month gap in the schedules before the next example (Chassis Number 3384FM) was built in April/May 1975. In the meantime, there had been a three-month hiatus following the factory fire in January 1975. Reintroduction of the 1600M had been planned towards the end of 1974, and the press release to that effect was circulated immediately *before* the fire.

The 1600M was brought back for 1975 because TVR had sensed a big change in their *home* market as a result of the Suez crisis and the petrol rationing scares of 1973–74; customers, it was thought, might want an alternative to the lusty and hairy-chested 3000M, especially if it could look just the same. The reintroduced 1600M, therefore, had the latest road wheels – 14in diameter, 6in rim width and a revised

M-Series chassis-frames ready for assembly at Bristol Avenue. The frame in the foreground is likely to form the basis of a 2500M, destined for the United States market. Some sort of tyre shortage was clearly in existence when this picture was taken, for several rolling chassis are being assembled on obsolete pressed-steel rims! (*Thoroughbred & Classic Cars*)

The fire which nearly killed-off TVR and put back the development of the company by at least a year, maybe more, started after business hours one day in January 1975 and consumed several finished and partly-complete TVRs. As firemen clear up the wreckage, a disconsolate Martin Lilley (in polo-neck sweater) looks on, before taking to his bed with double pneumonia, while Stan Kilcoyne (wearing the tie, in the background) looks even more depressed. Production began again inside a few weeks, but was not running at full-strength for another year or so. (*John Bailie*)

Psychologically, the most important date of 1975 was the one on which the first car to be built after the fire – a 3000M – was handed over to its new owner. The twin air intakes on the bonnet top were only a temporary design feature on this model. (*TVR*)

There were at least as many different facia styles as there were model years at TVR in the Seventies. This was the original 3000ML of October 1973, complete with walnut panel. (*John Bailie*)

styling pattern – along with the styling touches added to 2500M and 3000M cars in 1974. In the meantime, the price had rocketed – to £2,200 (basic) when reintroduced – and this increased progressively to £2,975 before the last car of all (Chassis Number 3938FM) was built in April 1977.

During the life of the car all the visual changes incorporated in the other M-Series cars were added, such as the deletion of the wing vents at the start of production in 1975, and the matt black sill and valance painting introduced on other cars in February 1974. In September 1976, for the 1977 model year, all the cars received a reshaped bonnet moulding, which incorporated an enlarged radiator air intake which also acted as something of a spoiler/stabilizer (this being more important for 3000Ms and Turbos than for the 1600M), and at the same time there was a revised facia, switch panel and tail-lamps. This was the second time that the tail-lamp style had been changed. All M-Series models had been treated to Triumph TR6-style clusters in 1973, in place of the original Cortina Mark II clusters, but this change had occurred when the 1600M was out of production.

Exactly 50 1600Ms were started in 1975, but this figure fell to a mere 29 in 1976 and only a single car was built and sold in 1977. At that time, the entire market profile for TVR had changed considerably, with attention becoming concentrated on the much more powerful cars using the 2,994cc Ford V6 engines.

The 2500M

Although the 2500M is a little-known model as far as British TVR enthusiasts are concerned, it was the most successful car ever built by the Blackpool-based company in the Seventies. Between 1972 and 1977, a total of 946 2500M production cars were built, and the vast majority of these were sent to the United States, where TVR Cars of America (led by Gerry Sagerman) found a ready sale for them.

The powertrain was that of the 'federalized' Triumph TR6, which is to say that the long-stroke 2,498cc straight-six-cylinder engine was fitted with twin Zenith-Stromberg carburettors instead of the Lucas fuel injection found on home-market TR6s. Peak power output varied slightly from

A raw M-Series bodyshell meets a slave chassis-frame for a fitting check before it goes into the paint shop to be sprayed, after which trimming and final assembly will follow. This was a typically busy Bristol Avenue scene in the Seventies. (*TVR*)

Chris Alford, genial Sales Director of John Britten Garages, TVR dealers in Arkley, near Barnet, giving demonstrations in his race-prepared 1600M at a Michelin International Car Test Day. Was his car wearing Michelins? It is doubtful. (*TVR*)

The original M-Series facia style was rather fussy and featured a padded dashboard... (*John Bailie*)

...but from the autumn of 1973 a new style was adopted, with a smarter steering wheel and yet more revisions to the heating and ventilation system. (*TVR*)

The original TVR body style was modified in 1960, at the time of the launch of the Grantura Mark II, giving the new car a considerably more balanced profile.

Prominent identification features of the body used for the Grantura Mark II are the flares over the rear wheelarches and the extended profile of the rear wings.

year to year, depending on the regulations and the amount of 'hang-on' gear required to satisfy the legislators, but Triumph usually quoted 106bhp (DIN) at 4,900rpm.

The engine was mated with the four-speed all-synchromesh Triumph gearbox (as used in the TR6 and the Stag, not to mention the 2000/2500 saloons and the Dolomite Sprint, though in the latter cases there were differences in detail), and Laycock overdrive was an optional extra and operated on top and third gears. Unlike the 1600M model, the 2500M used a Triumph TR6 differential and final-drive casing, so the chassis-frame of the TVR was modified to suit.

The first M-Series production-type chassis-frame (Chassis Number 2240TM) formed the basis of the first 2500M production car, work on which had begun in March 1972. Right from the start, the 2500M was intended to sell primarily in North America, though Martin Lilley also marketed it in the UK until May 1973. Its original UK basic price was £1,695 in June 1972, but this was raised to £1,779 in August 1972. However, as the 3000M was little more expensive and considerably faster than the 2500M, it overshadowed the latter completely. The consequence was that the 2500M was withdrawn from the home market and became TVR's USA-market car.

As an export car the 2500M was a great success and would undoubtedly have sold even more strongly had it not been marketed in harness with the old-style 2500 for the first year, and had not TVR sometimes found it so difficult to ensure supplies of engines, transmissions and final-drive assemblies due to the scandalous state of labour relations which persisted throughout British Leyland in the early and mid-Seventies. As it was, the total production of 2500s and 2500Ms was

This was the padded-facia style for the 2500M and 3000M models of the 1975–76 period; a major restyle was to follow in autumn 1976. (*John Bailie*)

1,334 in about seven years which, on its own, must have underpinned TVR's finances through a difficult period.

Production of 2500Ms rose to its height in 1974, when no fewer than 372 works orders were issued (compared with 49 orders issued for 3000Ms); because of the fire, and a lengthy strike at BL, this figure slumped to a mere 20 cars in 1975, which dealt a severe blow to TVR Cars of America – the sort of blow which afflicted many British-orientated businesses at this time.

The 2500M was effectively under sentence of death from 1974, when it became known that the Triumph TR6 was shortly to be supplanted by the TR7 (which had an entirely different four-cylinder engine). Production of the TR6, in fact, continued until July 1976, and the six-cylinder engine design remained in production (for the Triumph 2000 and 2500) until May 1977. Supplies to TVR then dried up rapidly, and the last 2500M of all (Chassis Number 4094TM) was built in July/August 1977.

During its five-year life, the 2500M received many significant styling and mechanical improvements, which were always made to the 1600M and 3000M models at the same time. These show that TVR were not only able to think and act very quickly because they were a small and flexible concern, but that Martin Lilley was always looking for some way to make his cars more appealing.

Visually, the external changes comprised the use of matt black front and rear valances and side sills, with the option of a vinyl-covered roof panel from February 1974, the addition of rubber instead of chrome facing to the bumpers from September 1974, the deletion of the front wing vents from the start-up of production after the fire in 1975 and the addition of the revised bonnet moulding for the 1977 model year cars. As far as the 2500M was concerned, however, the bonnet-top air intakes were always retained due to the tendency of the federalized TR6 engine (like most other de-toxed units with weak carburation) to run rather hot.

The only important chassis change was that the 5½J-15in light-alloy road wheels (of radial-spoke Vixen/Tuscan V6/2500 pattern) were abandoned in favour of 6JK-14in wheels of a different pattern in the autumn of 1973. At the

An early example of the 2500M rolling chassis, complete with Vixen-style cast-alloy road wheels, twin-Zenith-Stromberg carburettor, and USA-specification Triumph TR6 engine and gearbox. Overdrive was optional and – when fitted – was located behind the main gearbox casting and immediately under the gear-lever itself. (*Autocar*)

Like the Nineties variety, the original Griffith from the Sixties was a formidable machine, though it had to share the body style of the four-cylinder Granturas.

TVRs were still being built with a developed version of the Grantura-style bodyshell in the early Seventies, but were about to appear in the M-Type shape. This car's bonnet slots and bulges were typical of Tuscan V6s and 2500s of the period.

same time, the tyre size was changed from 165-15in to low-profile 185HR-14in.

Like other M-Series TVRs, the 2500M also received the periodic reshuffling in tail-lamp clusters and a quite confusing number of changes to the facia and instrument panel/switch/radio/heater control layouts. Early in 1974, too, better soundproofing, three additional stowage compartments, semi-reclining seats which could be folded forward to improve access to the loading area and other details were added.

In general, of course, it should be pointed out that many details of the fittings to US-market 2500Ms were different from those fitted to UK and European 1600Ms and 3000Ms, due to the very stringent and idiosyncratic regulations which applied in North America throughout the car's life.

The 3000M

The third of the M-Series models to be revealed publicly in June 1972 was the 3000M, though the first production car was not started until August. Without any doubt, this was the most popular UK-market TVR model during the Seventies. Indeed, from July 1973, after the elimination of the 1600M, TVR announced that UK and European markets would be covered by the 3000Ms, while the North American market would be covered solely by the 2500M, all of which was highly logical – but not for long.

In the case of the 3000M, the M-Series rolling chassis with its sleek two-door coupe body was powered by the 2,994cc Ford 'Essex' V6 engine, as had been fitted to the Tuscan V6 of 1969–71. This engine, along with the four-speed all-synchromesh manual gearbox mated to it, was normally fitted to 3-litre Capris and Granadas, and was also in use in other specialist sportscars competing with TVR, such as the Reliant Scimitar GTE and (to a lesser extent) the Gilbern Invader. As with the 2500M and earlier 2500 models, the final-drive was that of the Triumph TR6 (from early 1977, however, the TR6 final-drive was abandoned and its place taken by the Salisbury 4HU unit as used by the independent-suspension Jaguars of the Seventies).

So that costs would be minimized and service problems reduced, TVR used essentially standard engines, rated at the same peak of 138bhp (DIN) as those fitted to Capris and Granadas. Over the years this engine proved to be remarkably durable and – in the case of the turbocharged models covered in the next chapter – receptive to drastic tuning measures.

Apart from the changes in chassis, style and trim specification given to the 1600M and 2500M models and already described, the 3000M also came in for a realignment in its price and specification in the autumn of 1974. At that time, fittings like the reclining seats, sunroof, tinted glass, radio and heated rear window were all demoted from standard to optional extra equipment and the price was reduced by £235.

Two other important derivatives of the 3000M design and one 'Special Edition' should also be mentioned. From the autumn of 1975 (the feature was actually announced at the 1975 Earls Court motor show), the 3000M became available with the option of Laycock overdrive, which operated on top and third gears. It had always irritated specialist manufacturers like TVR that Ford never offered an overdrive on the Capris or Granadas using this gearbox. TVR solved it in their own way, building on the method evolved by Reliant for the Scimitar GTE.

Earlier, at the Earls Court motor show in October 1973, TVR showed off the 3000ML coupe, where 'L' stood for 'Luxury', and it denoted the use of an entirely changed, super-luxurious trim pack, where the facia was in wood (instead of being padded vinyl) and trim, carpets and seats were all of more expensive and exclusive specifications. However, this offering lasted for only one year and had gone by the end of 1974.

The 'Special Edition' 3000M was the 'Martin' model of summer/autumn 1976. Only 10 cars of this type were built, and they were conceived to commemorate the 10th anniversary of TVR production by TVR Engineering, the concern set up by Arthur and Martin Lilley at the end of 1965. The specification of these cars included Wolfrace road wheels, tinted glass, overdrive as standard equipment, sunroof, cassette radio and distinctive paintwork which incorporated the name 'Martin' picked out on a contrasting band of paintwork, along the flanks.

Even though the 3000M gave rise to the Taimar and – less directly – the Convertible, not forgetting the exciting Turbo derivatives, it outlived both its other M-Series relatives by more than two years. The last 3000M was Chassis Number 4940FM, started in November 1979, and in more than seven years a total of 654 cars were built. The most prolific year for the 3000M was 1976, when no fewer than 161 examples were started, even beating the 2500M, whose export success was beginning to run down a little.

The life of the 3000M was prolonged successfully by a very happy series of events. In 1974 and 1975, TVR became increasingly aware that their 2500M model was beginning to live on borrowed time, as its de-toxed Triumph engine would be phased-out of production relatively soon. Stewart Halstead, who became TVR's Sales Director in 1975, remembers this as an immediate and pressing worry: "In 1976 we flew out a car to Olsen Laboratories, in California. Olsen specialized in making engines 'clean' for California, where the regulations were even stricter than for the rest of the United States. We were the first manufacturers – as opposed to individuals – to go to him after he had approached us and claimed to be able to 'clean up' the Ford V6 unit." (At this time, incidentally,

TVR's most successful works-assisted driver in Prodsports racing was Colin Blower, here seen in the lead of a group which included a Morgan Plus-8, Stewart Halstead's TVR Convertible and a couple of Lotus models. (*Roger C Standish*)

This Griffith 400, registered in 1967 – one of the rare 'Martin Lilley' variety of V8-engined TVRs – reveals one way of cooling the big Ford-USA engine after a fast run!

This Griffith was built during the period when TVR were using Ford Cortina Mk I-style tail-lamps. Lightly modified by a later owner, this car reveals that although the short-wheelbase late-Sixties shell looked smart and compact, it still offered no external access to the boot stowage area. Hidden behind the impressive nose (right) is a lot of Ford-USA power and torque.

the British Ford 3-litre V6 had never been de-toxed by Ford for sale in North America, and there were no plans to do so.)

In 10 months, from start to finish, and including carrying out a 4,000-mile Assigned Deterioration Factor road test in California, Olsen made the V6 engine 'clean' for North America. "We started to send the first 3-litre 3000Ms to North America in January 1977, and it carried on successfully until the last of the M-Series cars were built at the end of 1979."

Whatever Olsen's demon tweaks were, and however they were achieved, need not be of concern in detail here. However, there must only have been a marginal reduction in power and torque compared with the UK/European engine specification, as *Road & Track* performance figures obtained in their December 1978 issue (actually for a Convertible) were almost identical to those obtained by *Autocar* for the Taimar hatchback in the UK.

Even by 1975, however, Martin Lilley's and Stewart Halstead's minds were on other matters. They were looking ahead to modified versions of the cars which had carried TVR so successfully from 1972 to the present. Not only were they now ready to evolve the hatchback and convertible versions of the cars which enthusiasts had been requesting for such a long time, but they were also ready to build a turbocharged car. All these models are so important to the evolving story that they deserve a chapter of their own.

No-one could miss the fact that this is a racing TVR. Stewart Halstead, TVR's Sales Director, used 'Martin's Mild Mooover' – a 3000M – with considerable success on the British Prodsports scene in the late Seventies. The scene is the chicane at Croft – you can't get much closer to the circuit wall than that! (*Tony Todd*)

Trying really hard at Silverstone, Colin Blower gets the Team Burgess 3000M well sideways when trying to keep ahead of a Morgan. (*Colin Blower*)

TVR's most popular model of the Seventies was the 3000M, of which 654 were built from 1972 to 1979. The hatchback Taimar was a close relative of this model.

Enthusiastic members of the TVR Car Club assembled their cars at a venue in Kent for this picture, which appeared on the front cover of the original edition of this Collector's Guide. The Tuscan represents one of the last developments of the original Grantura body style, the hatchback Taimar is a manifestation of the interim M-series body, and the Tasmin offers a preliminary look at the first range of cars featured in the companion to this book, *TVRs Volume 2: Tasmin to Chimaera.*

Footnote: Important modification points for M-Series cars (1972–79)

Date	Chassis Number	Detail
Mar 1972	2240TM	First M-Series chassis built
Aug 1972	2413FM	All subsequent cars to have servo-assisted brakes
Sept 1972	2451FM	All cars to have bonnet scoops deleted All cars to have larger lower grilles
June 1973	2767FM	All 3000Ms to have 2500M road springs
Sept 1973	2851FM	New-type steering wheel fitted
Dec 1973	2951FM	1974 specification began, new seats, etc
May 1974	3096TM	Start of fitment of rubber bumpers
Apr 1975	3380FM	First 'post-fire' car of 1975
Nov 1975	3502FM	1976 specification began
Mar 1977	3919FM	First cars fitted with Salisbury 4HU final-drives. A few cars subsequently without 4HU axle, but fitment continuous from 3955FM
July 1978	4425FM	Start of upright radiator for Taimars
Nov 1978	4542FM	Start of upright radiator on 3000M
Nov 1979	4699FM	Last odd-number in this series
	4970FM	Last even-number in this series

M-Series cars in motor racing

Modern motor racing is a specialized business and International events can only be won by freakish machines. Martin Lilley was far too wise to repeat the mistakes of 1962, but he was very happy to see TVRs winning in events where they could be competitive. Starting in 1975, therefore, he sponsored a small number of M-Series derivatives in the British 'Prodsports' series, where cars were not grouped by engine size, but by price. Martin had always been convinced that TVRs offered outstanding value for money, and victory in these events surely proved it.

TVR's involvement in Prodsports started with Colin Blower, who had written-off his Lotus Europa in 1974, met Martin Lilley, persuaded him to loan a 3000M for three events, and then tackled the 1975 series in a Burgess-sponsored car built by the factory. With one outright win and a number of top places that year, things were clearly promising, so for 1976 Team Burgess ran two 3000Ms, the other being driven by Rod Gretton. At the same time, TVR's Sales Director, Steward Halstead, joined in with another works 3000M and between them they recorded several outright wins in BRSCC and BRDC Prodsports Championship events.

In the meantime, Chris Alford, who was Sales Manager of John Britten Garages, of Arkley, North London (TVR main dealers), had started to used a 1600M in the same series, which found itself in another favourable price class and also won many rounds. That car, incidentally, was eventually sold to North Countryman Chris Meek, who campaigned it in Prodsports racing in 1979 with such success that he won the CAV-BRSCC series outright on points scoring.

With strong competition from cars as diverse as Morgan Plus-8s, Jaguar V12 E-types, Lotus Europas and even De Tomaso Panteras, things became really close over the next couple of years, with John Kent joining in as a works-assisted driver in his 3000M and Stewart Halstead changing to a Convertible in 1978. In 1979, everything began to come right for Colin Blower, who won many races, but finished second to Charles Morgan's Morgan Plus-8 in the CAV-BRSCC series and second in his class in the Triple-C Prodsports series.

By 1980, Stewart Halstead was really far too busy to continue motor racing seriously ("Besides," he told me, "Martin wanted me alive as a Sales Director, rather than as a dead hero!"), which was probably just as well, for Colin Blower wrote-off his Indestructible Socks-sponsored Convertible at Mallory in an early-season shunt and was very lucky to be able to take over Halstead's car for the rest of the season.

"After that," Colin says, "everything seemed to go our way. We won 22 races outright – every race we tackled in that car – which meant that we won the BRDC Prodsports series outright. At first we thought we'd won the CAV-BRSCC series as well, but after a lot of protesting from other people, Chris Meek was named the winner in his Panther Lima."

The M-Series ended there, in triumph, and (as far as TVR Engineering was concerned) at a very reasonable cost.

CHAPTER 6

Turbo, Taimar and Convertible

1975 to 1980

All three models surveyed in this chapter – the cars which made all the headlines for TVR in the late Seventies – evolved directly from the 3000M. In 1976–77 there was a sea-change in TVR marketing policy. Previously, one body style had been used, with a choice of engines; from mid-1977 this would be reversed, with one basic engine powering a choice of different body styles. It was one which made more visual impact on the market, even though TVR were not having to scratch around to keep themselves busy.

Chronologically, the Turbo came first, the Taimar followed and the Convertible was last of all – at least, as far as the public was concerned. Inside TVR, the idea of the Taimar was born first and there had always been the desire to build a convertible, but the decision to build the Turbo was taken abruptly, only weeks before it was revealed. Since the Turbo became a mechanical option available on *all* TVR models, it will be considered last of all. Firstly, the body changes will be explained.

In a way, it would be true to say that TVR and Martin Lilley were bullied into developing a hatchback model – the Taimar. Not only had such a versatile body style already been developed by their rivals – the Reliant Scimitar GTE in 1968, the Volvo 1800ES in 1971 and the Ford Capri II in 1974 – but the influential magazine road testers had begun to nag about it.

It was not that TVR had never wanted to do a hatchback, but that somehow they had never got round to developing the feature. At the beginning of the Seventies the SM, or Zante, prototype confused the issue for a time and immediately afterwards there was the considerable upheaval – commercial *and* financial – brought on by the introduction of the M-Series cars. In 1973 and 1974, following the UK's change-over from a Purchase Tax to a VAT system, almost every car which TVR could make was going to the United States. However, even at the British Earls Court motor show of 1973, TVR were telling motoring writers that a hatchback car was under consideration; Stewart Halstead remembers that when he joined TVR in the autumn of 1975 mock-ups had already been built.

In the event, the launch of the hatchback TVR, dubbed a Taimar (the name, incidentally, is not *quite* that of the river which divides Devon from Cornwall), was delayed until the Earls Court motor show in 1976. It was, in fact, something of a last-minute launch, for the style changes for 1977 models (described in Chapter 5) were introduced in September without mention of a hatchback derivative and the hatchback release was held back until the eve of the show itself. A study of factory records shows that the first *production* Taimar was not finished until November 1976 (after the show had closed) and was Chassis Number 3838FM. The car on show was actually the second prototype, completed in 1976 – the first having been made in 1974.

Mechanically, the Taimar was exactly like the 3000M, for differences were confined to the provision of the hatchback itself. With the back closed down, the two cars were virtually indistinguishable except for the panel joint line running around the lifting panel/window and on to the lip above the tail itself. There was no external release for the hatch – a release button was very neatly and cleverly positioned above

The feature most often criticized by testers of TVRs was the poor access to the luggage space. This was finally rectified in 1976, when the hatchback Taimar was announced. With the hatch closed the lines of the M-Series car were virtually unaltered, but with the hatch open there was no missing the new feature. Ahead of the doors there was absolutely no change to chassis or body, and behind that line the changes were kept to a minimum. (*TVR*)

The final coupe facia style of the Seventies (but not the final style of all, which was reserved for the Convertible of 1978–79) was introduced on 1977 model-year cars and included yet another relocation of minor instruments, heater controls and stowage space. This is a 1978-model Taimar facia. (*Autocar*)

the door lock in the driver's door jamb, where it could only be operated with the door open. It followed that when the door was closed the release button could not be pressed; the operation was electric, with a solenoid operating the catch at the rear of the gate. A nice touch was that this solenoid could only be 'live' when the driver's door was open, which helped to make the car more burglar-proof.

The hatch itself was bulky rather than heavy (the whole car,

The immensely practical hatchback feature of the Taimar is clearly shown in this studio shot. It also shows the final type of tail-lamp cluster adopted with the restyle of autumn 1976, which included yet another new arrangement of instruments and controls at the facia. When open, the hatch was held up by two gas-filled struts, and the release lock was actuated by a button hidden in one of the door jambs. (*John Bailie*)

With the hatch closed, there was virtually no evidence of the existence of this new body feature on the Taimar. In this view, the only real clue is that there are shut lines at each side of the tail, above the tail-lamps. The Taimar could be bought with the normally-aspirated (3000M) type of engine, or with the turbo-charged engine. (*John Bailie*)

in fact, was about 20lb heavier than the 3000M), hinged at the front and supported, when open, by gas-filled struts. It did not open perhaps as far as expected, merely because it would then have been impossible for a small driver to reach the sill to tug it back down again! Once open, of course, it gave unrivalled access to the loading area, which was wide and flat, though any contents stored there were still exposed to view through the huge rear window.

Only 15 Taimars were started in 1976, but a further 155 cars followed in 1977, which was more than the number of 3000Ms started (132) and instantly made the Taimar the most popular TVR in production. Once launched, the new TVR quickly became an established part of the scene at Blackpool. Not only did it sell well in the UK, but it also did well in North America, and the very last Taimars were built among the final batch of cars based on the M-Series chassis at the end of 1979. Altogether, 395 normally-aspirated

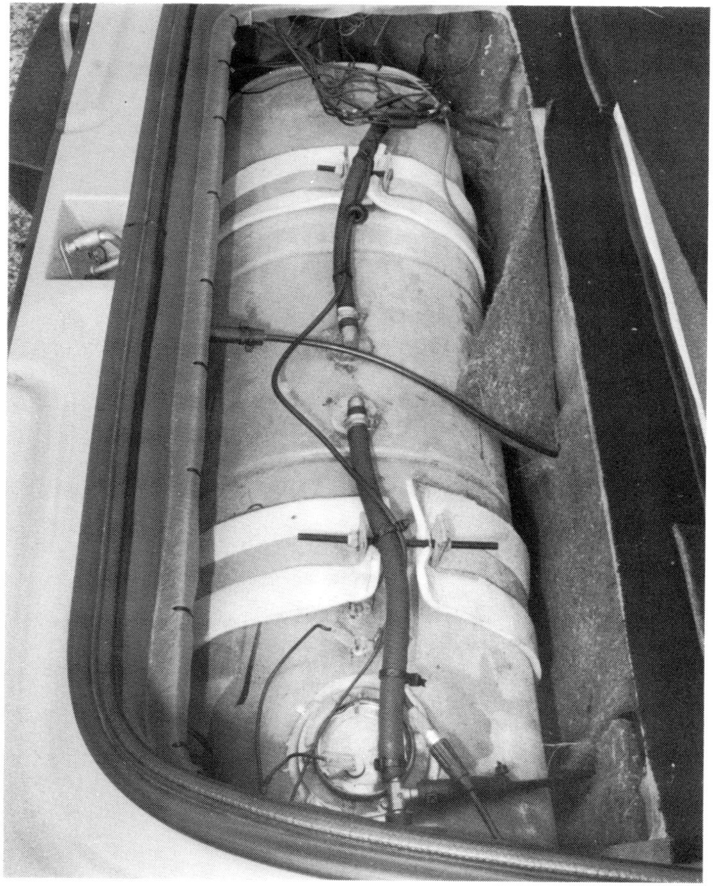

The fuel tank on all M-Series cars built from 1972 to 1979 was cylindrical in section, and was mounted across the tail of the car, low down and behind the line of the rear suspension. Access for service or repairs, or to reveal the electrical wiring leading to the tail-lamps, was gained by removing a panel in the loading floor. (*Autocar*)

Taimars were built in three years.

The other equally successful new TVR body style of the late Seventies was the Convertible. It is important to reiterate that this model was known merely as the Convertible – not the 3000M Convertible – for it had its own unique character and many special features in its construction. For all the same reasons as advanced for the late arrival of the Taimar, it took a long time to launch the Convertible, but there was another very important consideration – that of the rules applying to cars being sold in the United States.

In the Seventies, many manufacturers, especially those with limited resources, had to trim their ambitions to the amount of legislation with which they could deal, in the UK, Europe and especially in the United States. In the early Seventies, proposed US legislation led manufacturers as large and experienced as Datsun, Triumph and Fiat to assume that convertible cars would soon be banned from the new-car market place – the result being that the 240Z, the TR7 and the X1/9 and Lancia Beta Montecarlo models all had fixed-head styles. In the same way, TVR, who might have been tempted to produce a convertible model, dropped that idea when they saw the smoke signals. It was not until at least the

To celebrate 10 successful years of TVR production by TVR Engineering, 10 special 'Anniversary' cars, badged as 'Martins', were built in 1976. Among the obvious features were the Wolfrace wheels, tinted glass and the badging on the sides. (*TVR*)

mid-Seventies, when it became clear that such legislation would *not* now be enforced, that interest in convertibles was rekindled. TVR actually had their car ready for the beginning of 1978, ahead of Triumph (who similarly had second thoughts regarding the TR7, though in fairness the lengthy strike at Speke should be quoted as one reason why the drophead TR7 was not launched in 1978).

However, TVR's was no ordinary convertible. *Road & Track*, in their road test of December 1978 (when, quite wrongly, they called the car a Taimar Convertible), dreamed up this imaginary conversation at Blackpool:

"Okay chaps. Everyone else is dropping convertibles. So let's build one."

"Sure, but we couldn't design just another convertible."

"I've got it, mate. We'll build a convertible and give it side curtains"

– which sums up the design effort in only a few words. The Convertible was not merely a 3000M with the top chopped off, but a different and very genuine attempt to apply all the traditional British open-air virtues to a modern sportscar.

By comparison with the conversion from 3000M to Taimar, the production of the Convertible body was a much more complex job. As with the conversion of the 3000M to Taimar specification, no chassis changes were needed. Spring and damper settings were not altered, nor was the gearing, and the fuel tank remained in its usual place, behind the rear suspension, low down so as not to intrude into the space allocated for stowage. All Convertibles, however, used a Salisbury 4HU final-drive and differential. This feature had been standardized early in 1977 for two reasons: firstly, supplies of the Triumph TR6 final-drive were about to dry-up following the withdrawal of that car from production in 1976; secondly, the new final-drive was even more robust and would be able to look after the increased torque produced by turbocharged engines and by the fuel-injected German Ford V6 2.8-litre engine which was already known to be on the way.

Much of the Convertible's bodywork, however, was different. The easiest way to summarize this is to say that, visually at least, the changes start at the screen and work

Arthur Lilley and his son Martin – at that time Chairman and Managing Director of TVR Engineering Ltd, respectively – posing in front of one of the very first 3000M Turbos in 1976. It was the senior Mr Lilley's vision and – eventually – his son's application, which ensured the prosperity of TVR and its 100-strong staff throughout the Seventies and into the Eighties. (*TVR*)

backwards. The massive lift-up bonnet moulding was the same, as was the front and centre of the floorpan, but almost everything else was different. There were special doors, a special screen, an entirely different rear-end (this was the first and, so far, the only TVR to have a separate boot compartment) and a foldaway hood. Not only that, but the winding windows of the 3000M/Taimar models had been discarded in favour of glass sliding sidescreens which could be removed completely.

The style was effectively shaped by the doors. Their tops were carved out so as to allow arms to be extended into the (hopefully) warm breeze, and there was no window-winding

The 1978-model Taimar, revealing the final development of the M-Series frontal styling, which was actually introduced for 1977 model-year cars. Without the distinguishing stripe along the side of the car, it would be almost impossible to pick a 3000M from a Taimar at this angle. (*TVR*)

A fabulous 3000M Turbo, equipped with Wolfrace wheels. (*TVR*)

Hidden away under this bonnetful of machinery is the excellent Broadspeed turbocharged Ford 3-litre V6 engine, rated at 230bhp (DIN) and using a maximum boost of 9psi. The Broadspeed conversion had not been developed specially for TVR, but had first been seen on Broadspeed cars in the early Seventies. The turbocharger is situated, out of sight, below and ahead of the engine. (*TVR*)

mechanism. The glass windows were fixed on with two turn buckles, and when removed they had to be stowed in the boot; on early cars there was no stowage bag, an omission rectified on later examples.

The windscreen itself was no longer the familiar Consul style, but was new and special, with a bright metal surround. The hood itself was a genuine 'convertible', as opposed to a 'tourer' fitting, which means that it was permanently fixed to the sticks. Erection took less than a minute, and the car was reasonably windproof when all the fixings were applied.

Behind the seats there was the small, but versatile stowage area. As previously mentioned, it was the only TVR built up to that time in which there was a separate boot compartment, and one in which the items kept there could not be inspected by other people. There was also a panel at the front of this boot, which could be removed so that more bulky or awkwardly shaped packages could be installed and allowed to stick forward into the passenger compartment behind the seats.

Apart from the obvious increase in wind noise with the hood erect, and the greater difficulty in keeping the passengers warm in inclement weather, the only noticeable drawback of the Convertible was that it was rather less aerodynamically styled than the 3000M or the Taimar, and therefore a little slower and less economical. None of this seemed to harm its attraction at all to customers, for a total of 258 cars with the normal 138bhp Ford engine were built in less than two years. The last Convertible was also the last normally-aspirated M-Series car to be built (Chassis Number 4968FM) in November/December 1979.

Mention has not yet been made about the most controversial feature of the Convertible, and that which brought most criticism from road-testers: the facia layout. Although a change was somewhat forced on TVR by the fact that they were lowering the base of the windscreen (the 'cowl', as the Americans call it) by a couple of inches, it was a great pity that the position of the instruments was altered so much. TVR called their new layout 'traditional instrumentation', which meant whatever you wanted it to mean. The plain fact was that the speedometer and rev-counter could no longer be fitted between the crash roll over the facia and the steering column switch gear; accordingly, they were moved to

Part of the pipework involved in the installation of the turbocharger. This is an experimental set-up, but the principle was carried forward to production TVR Turbos. The turbocharger itself was positioned ahead of the main mass of the engine, and the reversed exhaust manifolds fed their gases forward to it to provide power to the spinning turbine. The black-painted 'horn' is part of the pipework normally connected to the pressurized plenum chamber above the V6 engine's Weber carburettor. (*TVR*)

new positions nearer the centre of the panel, widely separated, where they could not easily be read by the driver. Although the rev-counter could be considered to be the most important instrument, on left-hand-drive cars this was now completely remote from the driver's vision. *Road & Track* commented that "the tack is so far away you can't read anything below 4,000rpm ..." Speedometer and rev-counter positions, however, could be reversed if the customer requested this.

It is interesting to analyze the TVR sales story of normally

Turbocharged TVRs were very rare – only 63 were built between 1975 and 1979 – mainly because they were much more expensive than normally-aspirated versions. Here is one of them, in rolling-chassis form, during assembly at Blackpool. The cylindrical black object near the chassis crossmember is an air-cleaner for the turbocharger intake, and the turbocharger itself is below it, hidden under the light-coloured panel ahead of the crossmember which protects the spare wheel from the heat. In the background is a mixture of 3000Ms intended for different markets, the picture being taken before the Taimar or the Convertible had been introduced. (*TVR*)

aspirated-engine cars in 1977, 1978 and 1979: in 1977, works orders were for 155 Taimars, 132 3000Ms, 76 2500Ms and a single 1600M (the last of all); in 1978, works orders were for 129 Convertibles, 125 Taimars and 42 3000Ms; in 1979, works orders were for 129 Convertibles, 100 Taimars and 38 3000Ms.

On that evidence, there is no doubt that the Convertible was the most popular TVR of its day, or that it was the sort of open-air hairy-chested sportscar which the customers were looking to buy. All this was in spite of a US price of $15,900 for the first cars and $16,900 in 1979, and a UK basic price of £5,462 in 1978, which rose to £7,007 (total price £8,730) by the time the model was discontinued in 1980. Times and prices, indeed, had changed.

However, it would be true, if somewhat unfair, to point out that many of TVR's achievements in the late Seventies were completely overshadowed by the launching of the fabulous Turbo model. In quantity terms, the Turbo made little

Last, but by no means least, in the M-Series scheme of things was the Convertible, or 3000S as it was occasionally known, which TVR only built in 1978 and 1979. The rolling chassis was the same as that used on other M-Series cars, as was the front of the bodyshell, but the scuttle, windscreen, doors and entire rear-end were all unique to the Convertible. The side windows, incidentally, slid open and could not be retracted; instead, they had to be removed altogether. (*TVR*)

This 1979-model Convertible had been posed on a desolate spot overlooking Blackpool sands, on a rainy morning in April. However, with the side curtains fixed in place and the folding hood erect, the two lucky occupants would have been snug enough. The Wolfrace wheels fitted to this car were optional extras. (*John Bailie*)

Something very strange is evident here. TVR 22 is a registration number usually to be found on one of the TVR demonstrator car fleet, and in most respects the Convertible looks standard enough, but the facia and instrument layout is certainly not that normally fitted to production cars nor, indeed, to the Taimars and 3000Ms of the period. However, as has already been made clear, TVR *liked* to change their facias quite often ... In any case, this was only an experimental car, for there is no evidence of the opening boot either, with which all Convertibles actually sold were equipped. (*TVR*)

impression – only 63 cars of all body types were built between 1975 and 1979, which represents less than 5 per cent of production in the period – but it generated so much publicity and produced such startling performance figures for the technical press that it served its purpose as a company flagship. Whether it was profitable or not, nobody knows, but it was certainly a superb image-builder, and one that no-one at TVR regrets having put on the market. In its own way, the Turbo must have demonstrated that the TVR chassis was capable of handling much more power and torque than was normally asked of it, and in this way it must have generated additional sales of the normally-aspirated models.

As far as TVR themselves were concerned, the decision to make a turbocharged engine optionally available was taken during 1975, and the prototype was built in something of a rush so that it could be shown at the Earls Court motor show; that prototype, incidentally, was Chassis Number 3463FM, and soon after the show it was severely damaged in a road crash as development was being concluded. The engine conversion, however, had been in development for much longer than that, and this is where the story really starts.

The early examples of Ford's 'Essex' V6 engine, designed and built in the UK, were neither very powerful, nor noted for their rigidity and long life. That, however, was in the Sixties, and by the early Seventies they had been refined, made more durable and given revised cylinder heads, which released significantly more power. Ralph Broad, whose Broadspeed concern had been concentrating on the super-tuning of Ford engines since the late Sixties, turned his attention to this engine and began to develop a turbocharged unit. By the beginning of 1973, Broadspeed were demonstrating a turbocharged 3-litre Ford Capri and in the summer of that year (just before the Suez war and the energy crisis threw all such conversions back into the melting pot!) they also announced a Turbo Granada. In each case they were claiming 218bhp at 5,500rpm, with a peak torque of 225lb.ft at 3,500rpm.

In the end, Broadspeed never went ahead with the production of their own turbocharged cars (and after seeing TVR's own sales of such expensive and sophisticated machines it is easy to see why).

However, refinement of the system was carried out, and Martin Lilley finally linked up with Broadspeed, so that their engine could be used in the TVR chassis.

The engine as developed for a TVR was an extensively modified version of the 2,994cc British Ford 'Essex' V6 which TVR had been using since 1969, and TVR proposed to link it with an unmodified Granada/Capri 3-litre four-speed gearbox. At the rear, however, it was thought wise to specify the big Salisbury 4HU final-drive/differential and a slightly higher final-drive ratio than normal (3.31:1 instead of 3.45:1), but surprisingly a limited-slip device as an extra.

The plumbing of the engine was so complex that the author can do no better than quote Gordon Bruce's comments in *Motor* of July 10, 1976:

"Tracing the ... pipery is a job for a snakes and ladders expert, but to give you a rough idea, reversed TVR exhaust manifolds link up ahead of the engine where the turbocharger is situated. The outgoing gases then pass under the sump where they received maximum and very necessary cooling."

Another Bruce comment which tells all is:

"Atop the Broadspeed-red rocker covers is the airtight box that keeps the carburettor pressurized; on its side, the SU-based recirculating valve that bypasses some of the flow back to the turbocharger at low speeds, keeping the impeller spinning and all but overcoming the lag inherent in turbocharged systems."

In this developed form, the peak power output was no less than 230bhp (DIN) at 5,500rpm (a 67 per cent increase over the standard engine), while peak torque was 273lb.ft at 3,500rpm, which was a 57 per cent improvement. To achieve this, and apart from the use of turbocharging boost at a maximum pressure of 9psi, the entire engine was 'blueprinted' and balanced before final assembly, while the compression ratio was reduced to 8:1. It was amazing, but demonstrably true, that this engine could meet any existing European exhaust-emission requirement. Probably it could also have been adjusted to meet North American standards, but TVR never attempted this. Naturally, to keep the engine oil sump temperatures in check, an oil-cooler was needed. Testers found that the turbocharger effectively did not begin to work until 2,700rpm, before which the car performed like a normally-aspirated M-Series TVR.

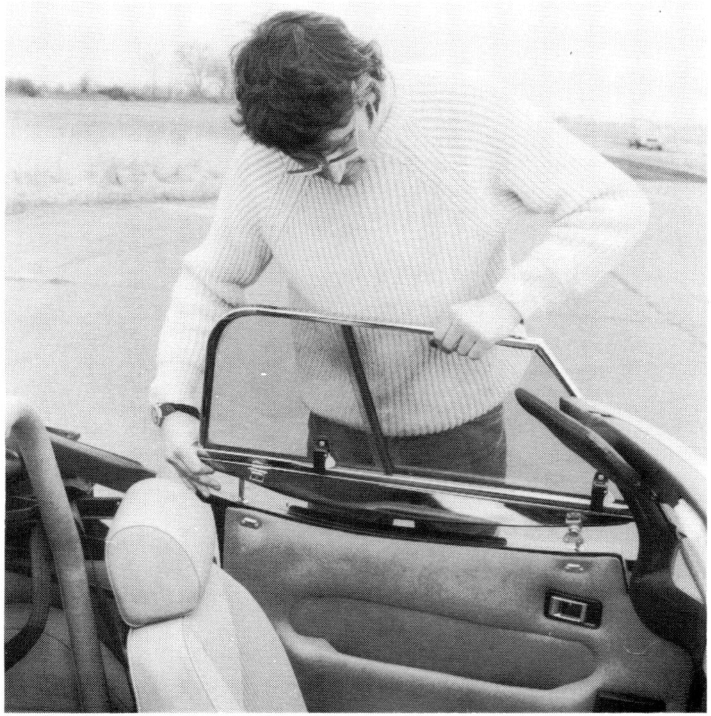

John Miles, of *Autocar*, demonstrating the way in which the glass sidescreens could be fixed or removed from the doors of the Convertible. Unlike all other TVRs built from 1958 to date, these side windows could not be wound down into the doors, but had to be taken off and stowed in the small luggage boot, or behind the seats. (*Autocar*)

Any TVR when fitted with this Turbo engine was remarkably fast. *Autocar*, who tried a Turbo Convertible in 1979, reminded their readers that its maximum speed (139mph) would undoubtedly have been higher if higher overall gearing had been specified and if the more aerodynamic coupe bodyshell had been provided. They were not, however, complaining about the acceleration, and made the telling point that the Turbo Convertible could sprint to a

Another *Autocar* illustration, showing the partly folded-down hood mechanism of the Convertible (in this case, *Autocar*'s car was a Turbocharged example). When the hood was folded back, there was no bag or cover to make the overall appearance completely tidy. There was extra access to the boot by a removable panel in the glassfibre bodyshell behind the seats. (*Autocar*)

quarter-mile from rest slightly faster than the exotic and much more costly Porsche 3-litre Turbo.

It was the sort of car which operated in superlatives, no matter how it was considered. Unfortunately, as far as sales were concerned, the same superlatives had to be applied to the customer's insurance costs, and to the extra price that had to be forked out for the privilege of being a Turbo owner. The Turbo derivatives were available on the 3000M right from the start, and on the Taimar and Convertible when they were announced. In the spring of 1976, when TVR were ready to deliver replicas of their own prototype, the extra cost was £2,490 (basic) and by 1980, when the last of the M-Series cars had just been built, that price had risen to £3,210. It was a credit to the Turbo, however, that when *Autocar* came to test the Turbo convertible, they compared it with cars like the Porsche 911s, 924 Turbos and Ferrari 308GTB two-seaters. Not only was it cheaper than any of these, but it also had better acceleration and a similar maximum speed.

Naturally, therefore, this was not a TVR for everybody, and the sales figures confirm this. With TVR Engineering's permission, the Chassis Numbers of every one of the 63 Turbos built between 1975 and 1979 have been listed, which show that there were 20 3000M Turbos, 30 Taimar Turbos and 13 Convertible Turbos. The most built in any one year was 22 cars – five 3000Ms, 10 Taimars and seven Convertibles – in 1978.

It was, however, a trailblazer in several ways. Not only was it TVR's first turbocharged model, but it was also the first British car to have turbocharging on a regular production basis. It also returned TVR to the standing of 'Supercar' builders, a position which they had reluctantly abandoned in 1970 when the last of the wide-bodied Tuscan V8s had left the factory.

In another way, too, it signalled the fact that TVR were now technically capable of tackling complex engineering jobs, which should have given their rivals cause to think about their intentions for the Eighties. For even while the Turbo TVRs were making all the headlines in the enthusiast press, TVR engineers were working on a brand-new car. The result was the Tasmin, which brought TVR smoothly and excitingly into the Eighties.

CHAPTER 7

TVRs that might have been

More prototypes and derivatives

Every car manufacturer has built prototypes which never saw the light of day, and most have initiated projects which never went much beyond the hopeful showing of a single car on motor show stands. TVR is no exception: there have been one-off cars, large and small, closed and open, which have been shown to enthusiasts, but were never put on sale. However, each of them, in one way or another, had an influence on the TVRs which actually were built.

In the beginning, of course, *every* TVR was a prototype and it was not until the end of the Fifties that anything approaching series production was achieved at Blackpool. Then, for a couple of years, the struggle to achieve profitability was so intense – and, in the end, unsuccessful – that there was little time to play about with new models. Before the free-spending era of the Aitchison-Hopton management began, the only real prototype to be considered by Trevor Wilkinson was a convertible version of the Grantura coupe, though one vain attempt at inserting a V8 2,548cc Daimler SP250 engine into the trailing-link chassis had also been made (that was a classic case of trying to get a quart into a pint pot and, like all such efforts, it failed).

Brian Hopton, Chairman and Managing Director of TVR Cars Ltd in 1961 and 1962, had many grandiose ideas, one of which was to put John Thurner's coil-spring chassis design into production under the existing Grantura bodyshell (this eventually took place during 1962) and, shortly afterwards, to discard the stumpy Grantura style and have it replaced by a new shell styled by an Italian specialist. This, however, remained a pipe-dream for some time as there was simply no money available to finance its construction.

There was also another Hopton-inspired project, designed by the aerodynamicist Frank Costin, which looked so repulsive that every true TVR enthusiast must be relieved that it was never put into production. Frank Costin was the ex-De Havilland engineer/aerodynamicist who had been involved with Jem Marsh in the evolution of the original Marcos cars, and was also well-known for his work in connection with the Grand Prix Vanwalls and Lotus 16 models. It was Costin, however, who had also been responsible for the design of a 4½-litre Le Mans Maserati racing sportscar, which was also about as ugly as they come.

At the time of its conception by Costin, who had then struck out on his own, the P5 was based around the concept of a rear-mounted engine and gearbox. The prototype, started in 1962, was powered by a three-cylinder German two-stroke DKW engine and transmission (the DKW, incidentally, was a front-engine/front-wheel-drive car), though it was thought that a TVR production car might be powered by the Rootes Imp engine/transmission, which was already something of an open secret in the motor industry and was scheduled to go into production early in 1963.

The chassis might have worked as an engineering layout, but the styling of the coupe was bulbous and nasty, for Frank Costin was not at all interested in looks, but *totally* interested in the drag coefficient of his cars. The screen was low, the door glasses were huge, the pillars were thick and the wheelarches partly faired-in. It was a monster, and deserved what it got – cancellation.

Brian Hopton's 'Italian-style' project came to nothing under his management, but it sprung back into life in 1963 after Trevor Fiore (actually a British North Countryman originally called Trevor Frost!) introduced himself to Bernard Williams and suggested that his own links with Fissore of Turin might be the ideal way of turning new styling ideas into reality. Money, however, was the reason why Fiore's plans could not immediately be carried out (money was *always* a problem at TVR in the Fifties and Sixties), the result being that his ideas for a new shape of TVR pre-dated those for the Elva-BMW style which was first shown to the public.

The radically new TVR style, therefore, was not actually carried out until 1964 (by which time the company operating at Blackpool was Grantura Engineering) and it was given the name Trident, a title originally coined for the nasty-looking Costin-designed P5. Originally, it was to have been built up on the basis of the 7ft 1.5in-wheelbase coil-sprung chassis, which was standard for both the existing Grantura Mark III and the Griffith 400, which dominated the scene at TVR for the moment. However, even though work had actually started at Fissore in Turin (who had, indeed, co-operated, for a fee, in the Fiore-Williams schemes), a request from Jack Griffith in North America for the new car to be powered by an American Ford V8 unit led to the wheelbase being lengthened to 7ft 6in, merely to make installation of the bulky American engine that much easier.

Two Trident prototypes – both sleek fastback coupes with light-alloy bodies (production cars, scheduled for mid-1965, though this target was completely impossible to meet, were to be in glassfibre) – were ready for the Geneva show in March 1965. One car was altered at Arnold Burton's insistence, after the styling had been completed, so that there was more interior headroom, and carrying this out involved raising the entire waistline of the delicately styled product; Trevor Fiore, reputedly, was not amused. That – as far as Fiore was concerned – was the bad news; the good news was that the original style, the one with which he was completely satisfied, was the TVR put on display on the Fiore-Fissore stand at the Geneva show.

Mechanically, the first Trident was really a long-wheelbase Griffith 400, but with four-wheel disc brakes. *Autocar*, which was still fighting its way out of a very stodgy editorial period, described the Trident as having 'quarried rather than sculptured lines', which was rather a sniffy way of pointing out that the Fiore style was a sharp-edged shape, rather than one with bulbous and flowing lines. They did, however, publish a detail illustration of the way the hidden headlamps popped up from the slopes of the bonnet for night-time use. Such a layout was normal on the Lotus Elan of the day, but on true 'Supercars', of which the Trident was a typical example, the idea was new. *Autocar* also made the point that the car displayed was mechanically incomplete. However, the second car, which had been back to Blackpool for finishing-off, had then been driven out to Geneva for use as a demonstration car.

The original concept sketch, by Trevor Fiore, for the TVR Trident coupe. *(John Bailie)*

After the Geneva show, the display car was shipped off to the New York show, in April, and two further prototypes were put in hand at Fissore, both with light-alloy bodies and one of them to be an open version. An attempt by Jack Griffith to have the Trident known as a Griffith was turned down, and the projected June 1965 introduction date for glassfibre-bodied cars was soon forgotten as Grantura

The first Trident prototype, constructed by Fissore, ready for display at the Geneva motor show of 1965. Underneath was a longer-wheelbase coil-sprung chassis of the type TVR used from 1962 to 1972. This prototype was built in steel and light-alloy, but the intention was that production cars should have a glassfibre bodyshell. (*John Bailie*)

Engineering slid inexorably towards bankruptcy.

The events which followed the closure of Grantura Engineering in the summer of 1965 were complicated. Although the Lilleys – Arthur and Martin – reputedly thought they were taking over the rights to the Trident when they bought the bankrupt company's assets, they were astonished to find that the TVR dealer from Suffolk – Bill Last (whose stand had hosted the British-market Griffith at the Racing Car Show of January 1965) – had bought the original Fissore-produced glassfibre body moulds for the Trident, and that in January 1966 he showed off a car at the Racing Car Show which he simply called a Trident. This had an Austin-Healey 3000 chassis, underframe and suspensions, but was quite clearly a convertible version of the original TVR Trident style. Certainly it was not as sleek, nor as distinguished as the 1965 TVR prototype because, once again, the wheelbase had had to be stretched to match the Big Healey's 7ft 8in dimension, there were big hot-air outlets behind the front wheels and the headlamps were now recessed into an altogether less delicately styled nose. The prototype, like those built for TVR, was equipped with a light-alloy bodyshell, though Last let it be known that production Tridents would have glassfibre shells.

The sale of the body style to Last had been encouraged by Trevor Fiore, who thought the rights were still his. However questionable this might have been, by the time the Lilleys got around to considering it, a *fait accompli* was obvious. Although TVR Engineering were in possession of several part-finished prototypes, their design had been 'acquired' by Bill Last. How, when, and in what detail the impasse was resolved is not completely known, nor does it really have any place in this book, for TVR's involvement in the Trident project ceased with the bankruptcy of Grantura Engineering. It goes without saying that Bill Last's presence in the list of TVR dealers came to an abrupt end, and it must have been a source of some satisfaction to Martin Lilley that the 'TVR that Got Away' was never more than a partial success for Trident Cars, no matter what engine was used to power it. It should be re-emphasized, if any casual TVR enthusiast does not already realize it, that there was absolutely no connection

Due to a mix-up of rights and trademarks following the financial collapse of Grantura Engineering in 1965, the Trident design fell into the hands of Bill Last, an erstwhile TVR dealer from East Anglia. This convertible was the first further derivative of the original Fiore-Fissore design, which he called a Trident and which featured an Austin-Healey 3000 chassis, a longer wheelbase and a different nose style. (*Thoroughbred & Classic Cars*)

The original TVR Tina of 1966, based on the floorpan, suspensions and powertrain of a Hillman Imp Sport. It was styled by Trevor Fiore, built by Fissore, and utilized steel for all its panels. The original droop-snoot body style was not liked and underwent revision before the project was shown again in the autumn of 1967. In addition to this convertible, there was also a fast-back hardtop derivative, but neither car was ever put into production due to the high capital cost likely to be incurred in tooling. (*Autocar*)

Two more detail views of the neat Hillman Imp-engined Tina. Had the car gone into production, was its shape too short-lived and 'fashionable' to have survived a long career? (*TVR*)

between TVR Engineering and Trident Cars.

Once Martin Lilley took up management control at Blackpool, he immediately started looking round for ways to update TVR's image and widen the appeal of the marque. The first way in which he tackled this was by consulting Trevor Fiore and the Fissore coachbuilding works in Italy over a new small car. There was something of a love affair between stylists and limited-production car engineers in Britain at the time, and the delicate – if flawed – engineering of the Rootes Imp engine and transmission, which was not only compact and light in weight, but had already proved to be eminently tunable for competition purposes. Martin Lilley and Trevor Fiore decided not only to use the Imp's power pack, but based their entire design on the floorpan and suspensions from the Imp saloon car. It was a time when such things were fashionable, the Fiat 850 Coupe and Spider models being the obvious market leaders.

TVR's basic mistake regarding the small car – dubbed a Tina – was made right at the start, at the design stage. It had nothing to do with the basic engineering of the car, nor with the looks of the prototype. It was simply that the steel floorpan was mated with a steel bodyshell. Unfortunately, this was not merely to get a complete car built, with a view to converting to glassfibre for the production cars; it was seriously thought that TVR could somehow find the money and facilities to have steel-bodied production cars built in reasonable quantities and be able to sell them at the right price. It was thought to be technically impossible to mate a glassfibre superstructure to the pressed-steel underpan (such matings, however, have since been achieved on other cars). Clearly, the Tina, having an 875cc Imp Sport engine, would have to undersell the Grantura Mark IVS, whose 1966 motor show price was £988 (basic) in kit form.

The original prototype, finished off in a great hurry for display by Fissore at the Turin motor show, in November 1966, was a rather plain little convertible with headlamps recessed in a drooping nose behind contoured Perspex covers. Other sources have suggested that there were family resemblances to Fiore's Trident style, which was wishful thinking as there was scant evidence of the sharp-edge style of

Having built the Tina roadster in 1966, TVR then built a fastback coupe, which was shown at the 1967 London motor show. It was still based on Hillman Imp engineering, but the nose style was radically different from that of the first roadster. For the author this car was as smart as anything being produced at the time by the so-called Italian 'experts'. (*TVR*)

the earlier car and the scale was completely different. As with all existing Imps, it showed off the strong positive camber of the wheels at the front and this, together with the rather anonymous front style, made it look more like an overgrown toy than a serious project. The author, too, can recall inspecting this original car in the company of a top Rootes engineering executive, who sadly pointed out the number of aspects which would need attention before it could be legally sold, even in Europe. The most obvious of these failings was the position of the headlamps, but there were many other details.

Once seen, the Tina disappeared from public view for another year, and it was not until TVR took a stand at Earls Court in 1967 that the British public had their first look at the project. By this time, two cars had been built – the original convertible, suitably modified in many ways to include a more bluff nose and exposed, rectangular headlamps, and a fastback coupe version of the design; both were put on display at Earls Court.

The problem, by then, was that Arthur and Martin Lilley had both become convinced that there was no way in which TVR Engineering could take on the complete manufacture of the Tina unless its all-steel body construction was changed. In spite of desultory talks with other manufacturers – TVR even tried to talk Rootes into building the car themselves, unaware, until informed, that Rootes were already studying such a project of their own design (this was why the Rootes engineer had been so knowledgeable when the Tina was discussed with him in 1966) – no advances were made and the Tina project was abandoned. The fastback coupe remained at the factory, while the convertible was acquired by a British enthusiast.

Apart from oddities produced on otherwise standard TVR chassis, like Tommy Entwistle's Gem and a number of self-modified specially styled cars produced by TVR dealers or customers, there is one other TVR prototype which should be mentioned, not only because of its startling shape, but also because of the time it appeared. The year was 1971 and the car was the SM – later known as the Zante.

Zante is the name of a tiny Greek island, but SM stands for Specialised Mouldings, of Huntingdon, who were already

The full-scale mock-up of the SM (or Zante) project of 1971. This was to have used the new M-Series chassis and a variety of engines, and would have been the first TVR to have an opening hatch to the loading area in the rear. In some ways, its frontal style was repeated by the Tasmin of 1980. (*John Bailie*)

For a time TVR was very serious about putting the SM/Zante model into production. After developing a full-scale mock-up, a running prototype was built and shown at Earls Court in 1971, but it was soon abandoned in favour of the M-Series models. (*TVR*)

famous for their bodywork fabrication for the racing car industry. On this occasion, the styling was not by Trevor Fiore (all links with him and Fissore were broken after the pretty Tina project was cancelled) but by a BL styling engineer. The author has seen similar shapes to the SM in the Rover-Triumph styling archive of the period...

The basis of the SM was a standard TVR chassis of the period, complete with the six-cylinder Triumph 2500 engine and transmission. The body itself was in glassfibre and was a two-seater like all previous TVRs. However, it conformed to almost every current trend, with its headlamps hidden away under flaps unless needed (like many modern Italian 'Supercars' and the original TVR Trident of 1965), a wedge front style, an upswept tail and a full-width combined lift-up tailgate and rear window like those of the Reliant Scimitar GTE or the then-new Volvo 1800ES. Brian Hatton, *Motor*'s styling specialist, loved it and commented: "With development, this car could well rival the appearance of many exotic continentals without involving the customer/builder in an exotic outlay."

Unhappily for TVR, who were heavily dependent on sales to Gerry Sagerman in the United States, Sagerman was not very impressed and thought it could not possibly have enough performance without a larger and more lusty engine (a big American V8, perhaps?). However, such engines had always been bad news for TVR in the past, and this tepid response, and the overall depression brought on by the burgeoning impact of new North American safety regulations, led Martin Lilley to drop the Zante project after a great deal of money had been spent and one complete car built. The car itself was eventually dismembered and the bodyshell brought back to Blackpool. Even as late as 1976, I saw the dusty remains of the Zante shell tucked into a corner of the factory at Bristol Avenue. From time to time, Stewart Halstead told me, thoughts had returned to the Zante and what might be done with it, but somehow work had never restarted; in the end it was abandoned, as thoughts of the new Tasmin took its place.

Even today, TVR is no doubt working on several new projects, for it is clearly not a hidebound, inward-looking concern which never makes plans for its future. Someday, perhaps, evidence will be seen of other cars which were designed but never put into production, since not all of them can come to fruition.

CHAPTER 8

Buying an older TVR

The choice, the examination and the road test

The author's aim in this chapter is merely to offer advice – hopefully presented in a logical manner – to anyone who might be considering the purchase of a car of the type covered. Such a car may be quite old, will almost certainly have dropped out of production and may have suffered neglect by one or more previous owners. There is, therefore, no foolproof way of offering advice which holds good for every car and every model considered. In the case of the TVR, advice is given as a result of the experience amassed with various cars, new or not so new.

The first and probably most important piece of advice to anyone considering buying an older TVR is that they should get a good idea of the cars' character when they were new. There is absolutely no point, for example, in trying out a bedraggled Grantura Mark II or Mark III and expecting it to be as quiet, well-finished or well-behaved as, say, an MGB or a Triumph TR4. Older TVRs simply weren't as 'ordinary' as that.

It is hoped, as a result of the many facts this book contains, that a good basic grounding will already have been gained by the reader. Much may also be derived by reading reprints or originals of the impartial and accurate road tests carried out by the more reliable British and North American motoring magazines. In that category would be included *Autocar*, *Motor* and *Autosport* from the UK, along with *Road & Track* and *Car and Driver* from North America. The British monthlies – *Thoroughbred & Classic Cars* and *Classic and Sportscar*, for example – will also provide a 'good read' and some background knowledge, but neither tend to give as much statistical and performance data as the more 'technical' publications.

The next point to be emphasized in connection with TVRs is that the supply of older models is very limited. Although the onset of terminal corrosion can be a problem, it is also a fact that the cars were more than usually vulnerable to crash damage. Therefore, although TVRs should not have suffered structural collapse due to body rusting – the glassfibre bodies cannot rust, though they can go brittle with old age and rough use, while the multi-tubular frames have often proved to be durable and long-living – many may have been scrapped after a serious accident. Leaning a TVR against a lorry, or the scenery, sometimes leads to shredding of bodywork and serious distortion of the frame and, as a result, many cars (intrinsically not worth much at the time of the accident) were scrapped there and then.

There is also the inescapable fact that it took until 1980 for the 5,000th TVR to be built and that before the Lilley family started to build TVRs in 1966 only about 1,000 cars had been built in eight years. As explained in Appendix D, no more than intelligent estimates of production and exports for the first 10 years of TVR's life can be given. It could be guessed, therefore, that about 60 per cent of all pre-Lilley TVRs were exported (almost all of the V8-engined Griffiths went to North America, for instance).

However, detail figures are available for the Seventies, and these show that of the 3,290 cars built, 1,740 (53 per cent) were exported, the majority of those going to TVR Cars of America. In some years, and in the case of some models like the 2500 and 2500M derivatives, the proportion was very much higher.

All of which means, inescapably, that the balance of

available TVRs leans very heavily to the more modern types – the M-Series and Tasmins. Perhaps the 'classic car' mania for collecting anything, and paying inflated prices for it, passed away with the Eighties, which means that asking prices for older (Sixties-built) TVRs may have returned to more sensible levels, even though there are relatively few left intact.

It is now time to risk nominating the 'Best Buys' and point out the less attractive propositions. What follows is, obviously, a *personal* opinion; it may not be yours (which means, of course, that some TVR sellers will be delighted to know it).

Straight away it must be said that the author is no great fan of the original-type 'trailing-arm chassis' TVRs. Not only was product quality often suspect – water leaks, wind-sealing, trim and bodywork finish defective – but the chassis was of a design which was neither very rigid, nor very long-lasting in retaining its settings. The suspension settings, when the cars were new and in 'as-designed' condition, gave a very hard ride. The ground clearance was limited, which could lead to the underside and the exhaust system suffering badly on unmade roads. As the suspension began to wear, play in the

A line-up of new M-Series bodies outside the moulding shop at Bristol Avenue, prior to painting. (*Thoroughbred & Classic Cars*)

Mounted on a slave chassis-frame, this M-Series bodyshell is about to be sent into the paint shop, followed by delivery to the main assembly lines in another building. The glassfibre bodyshells were really very simple, and many enthusiastic restorers should be able to achieve the same standard of finish on their cars. (*Thoroughbred & Classic Cars*)

trailing-arm pivots developed and the result was that the early Granturas showed a degree of 'self-steering' which could be alarming, if not downright dangerous. Oversteer was normal as a matter of course (the fact that the wheels rolled at the same rate as the chassis didn't help this), but once the wear set in this was intensified.

Nevertheless, the early 'trailing-arm-chassis' cars have a certain charm, if only because they *were*, after all, the original TVRs. They were, too, the lightest of all the models and a fair number were equipped with what was, for its day, the advanced, powerful and efficient single-overhead-camshaft Coventry Climax engine. The Climax-engined Granturas were quite fast – as the performance data in Appendix E makes clear – but on the other hand, those fitted with the small-capacity Ford engines were definitely qualified as sheep in wolves' clothing, for their performance was very limited indeed. It is difficult to understand why the builders of specialist cars were always so ready to use the side-valve 1,172cc Ford 100E engine at all, for its design was ancient (its general layout dated from the Thirties) and its power tuning was limited – it was, presumably, very cheap to buy from Ford! The overhead-valve 997cc Type 105E engine was nearly as gutless, though the 1,340cc Type 109E engine (with 54bhp) was better.

Even citizens of North America should be extremely cautious when considering the purchase of a Sixties-style Griffith, not because of its enormous performance, but because of all the known quality and reliability problems which were an unfortunate feature of these cars. Those which survive might well have been extensively improved by their previous owners, but as this might also mean that they are considerably 'non-standard' the potential buyer is in something of a cleft stick.

The buyer should also be somewhat chary of the attractions of a Turbo. The two major problems are that the engine is a highly-tuned and somewhat costly unit; and that the concern which built them has now gone into liquidation. If the engine should give trouble, therefore, the possibility of a large repair bill has to be faced (turbochargers do not come cheap, nor are specialists well-spread around the motor trade). TVR relied heavily on Broadspeed, of Southam, Warwickshire, for service expertise. That company closed down at the beginning of 1981. TVR themselves can now advise on the maintenance of the engines.

Experience with other marques and the whole historic car scene suggests that the Convertible – built only in 1978 and 1979 – will always be the most desirable of all 'classic' TVRs, merely *because* it was a convertible; the preservers of fine and interesting cars like to have an open-topped version if at all possible. However, the most practical of all TVRs was probably the Taimar, which was distinguished by the liftback 'third door', which gave unrivalled access to the rear loading area. Along with the Convertible, the Taimar was one of only two TVR production cars in which access to the luggage stowage area was really acceptable. On all other cars it required some physical strength and a degree of suppleness to get at the luggage from one or other of the passenger doors.

In this connection, attention should be drawn to the fact that no TVR built before 1980 had a sensibly-positioned spare wheel, so strong biceps had better be developed to

New TVR bodies being built in the Seventies in Blackpool. Because David Gerald TVR has purchased all the moulds, it is still possible for new shells – or parts of shells – to be manufactured today. (*TVR*)

extract it when the time comes! Taking the spare out of the front of the engine bay on M-Series cars is a dirty job, and a hernia could be incurred in heaving it out of the tail of the early Granturas; in all other cars it had to share space with the luggage behind the seats.

My favourite TVRs, therefore, judged not only on their performance, but on their practicality, are the 3-litre V6-engined cars, preferably the Taimars, but also the very numerous 3000M models. This is not to say that a healthy respect for the four-cylinder Granturas and Vixens is lacking; it is merely that the author likes torque delivered in large lumps and from low rpm. If a four-cylinder-engined car had to be chosen, it would probably be a Ford Cortina/Capri-engined Vixen, for these cars not only have free-revving engines which seem to last an astonishingly long time, but they also have splendid light-action gearboxes. All this, and Ford service, make them a more attractive proposition than the cars fitted with slower-revving MG engines and BMC gearboxes with 'crash' first gear which – somehow – had a little less 'verve'.

So far the Triumph-engined cars have barely been mentioned. The 1300 model can virtually be dismissed as only 15 of these cars were built and several of them will probably have been re-engined at some stage in their lives to give them more performance. The 2500s and 2500Ms, of course, have been the most popular TVRs in North America, which was precisely why they were developed and marketed in that continent, but they are extremely rare in the UK and Europe. The 2500M, in particular, is a fine car and probably even smoother than the other M-Series models, but it is by no means as fast – the fact that its engine was rated at 106bhp, compared with the 138bhp of the 3000M/Taimar, proves this conclusively. Nevertheless, a quarter of all TVRs built have had the straight-six Triumph engine fitted, which makes them the most numerous of all. If I lived in North America, I would still choose one of the 1978–79 cars with the 'federalized' Ford V6 engine – if one could be found – otherwise I would be happy to settle for the silky and refined behaviour of a well-adjusted 2500M.

It would be quite impossible to advise on the purchase of a particular car in print. However, it is worth pointing out that the oldest of the series-production TVRs is now well over 30 years old and that even the youngest of the pre-M-Series models is now past its 10th birthday. Even though all the cars should be reasonably corrosion-proof (the tubular frames seem to last for many years before becoming structurally defective, while the glassfibre bodies should be impervious to any sort of deterioration except that of becoming brittle in certain circumstances), they are all going to *feel* old and tired in due course, not least in the way the bodies fit together, repel the wind and rain and stay rigid and creak-free.

Restoring a TVR to its former – and proper – glory is not for the faint-hearted, but at least there should not be the heart-breaking task of getting rid of acres of rust, and there does seem to be a ready supply of spare parts, large or small, and the expertise to go with them; the final chapter covers this aspect of preservation in more detail.

The M-Series chassis-frame in course of construction. One of the main problems facing a TVR restorer may be that the frame has been distorted in a previous accident. It may be possible to jig-up the frame to make sure it is square in every respect. (*John Bailie*)

At this stage of rebuild every enthusiast's TVR rolling chassis should look as smart as this one. This was the state of a 2500M, as newly completed at the factory, in 1972 or 1973. The pattern of the cast-alloy wheels, which are the same as those used on late-model Tuscans and Vixens, confirm that date, and the engine, of course, is the US-specification Triumph TR6 unit. (*John Bailie*)

Even before buying, however, a thorough road test is essential. All TVRs – and the earlier models with the less rigid frames were worse in this respect than their successors – were vulnerable to the accident which distorted the chassis, but was not serious enough to write-off the car completely. On pre-1967 models, in any case, the frame was bonded to the bodyshell, which made separation of body from chassis, to aid repairs, quite impossible. Since it is difficult to straighten a multi-tubular frame absolutely correctly without a great deal of time spent on the job (and a jig), there are probably a number of TVRs running around where the alignment is not what it should be. One object of the road test, therefore, should be to assess the car's straight-running behaviour and its cornering abilities to right and left. Do not be too proud to demand to sit behind the car as it goes along, in a friend's car, so that its behaviour can be personally observed.

Pre-Lilley-management TVRs were not altogether renowned for the consistent quality of their construction (many of them, after all, were built by their first owners from kits), so always be on the look-out for evidence of water leaks around the doors and front and rear screens, and for draughts from badly matching door/bodyshell apertures. Do not expect the output of the heating and ventilation system to be satisfactory – even on a late-model M-Series car.

Incidentally, and this point is mentioned in other chapters, it is worth emphasizing that great difficulty may be experienced in finding an older TVR which has not been 'customized' or made non-standard in some respects. This is because the glassfibre bodies are so easily altered by relatively unskilled mechanics, at quite a low cost (such things could certainly not be tackled on a car having a pressed-steel or a hand-built bodyshell) and it may mean that 'your' car may have slots or scoops where none should exist, or standard air-intakes or outlets modified or blanked-off altogether. The heartening point to be made is that you, like the previous owner, can also change the contours fairly easily to restore the originality.

If you are still not sure you can properly 'vet' a TVR for originality, acquaint yourself with a club member who knows about these matters. There is nothing quite as impressive as the special knowledge held by an active member of the appropriate one-make club. The TVR Car Club is no exception – it ranks specialist 'historians' among its membership.

Much will eventually depend on your need for spare parts for the car you have decided to buy, and the situation as regards maintenance and restoration – issues to which the last chapter in this book is devoted. However, on the subject of restoration, it is surely not necessary to emphasize that dealing with the refurbishment, alteration and substitution of glassfibre panelling is not an activity for the novice. At certain stages of the game there may be fire risks and at others there may be a danger of inhaling the fibres themselves. While it is agreed that glassfibre is probably the easiest material for the ill-equipped amateur to attempt to repair without investing in expensive equipment, you are urged to take all the proper precautions before doing so. Useful advice of this nature will be found in Dennis Foy's book *Automotive Glassfibre* (MRP).

CHAPTER 9

Maintenance and fellowship

The Club, spares and restoration

In some previous *Collector's Guides*, dealing with cars which had been built in large quantities, but had usually dropped out of production, it has often been reported that the provision of spare parts and service expertise has been abandoned by the makers of that car, and that the specialist club has had to take over. In the case of TVR the situation is much more encouraging. Not only is TVR still very much in business, but a separate organization now looks after spare parts supply and the service to back them up has been maintained. Better still, not only is there a thriving TVR Car Club, but its relations with the factory are close and cordial.

It is much easier to find, and to identify, the spare parts needed for an older TVR than might be the case with a more common sportscar. Until the mid-Eighties this was due to TVR having a policy of keeping a spare parts capability for its older cars as long as there was space at Blackpool (and it made economic sense), but since 1987 it has been down to the enterprise of David Gerald TVR, which took over parts stocks, jigs, moulds and the rights to make more parts as required.

In preparing this chapter, I had long talks with people like Stan Kilcoyne, as well as officials and active members of the TVR Car Club. It is indicative of the close links between the Club and TVR Engineering Ltd that Peter Wheeler is the Club's President. For many years John Bailie – who prepared the splendid 3000M cutaway drawing which appears in Chapter 5 – was not only a Club committee member, but also sold his professional skills as a graphics and publicity material designer to the TVR management.

For the TVR enthusiast who wants to do his own research, Parts Manuals are available for all cars built since TVR Engineering Ltd was set up in 1965–66, which is to say that there is full and detailed coverage of Vixens, Tuscans and their derivatives, all the M-Series cars and, of course, the 3000S.

It is surely not necessary to point out that many proprietary components – those designed for mass-production BL, Ford and similar cars – have been used in all TVRs, the Ms and Taimars no less than the original Granturas. This means that it may be possible for items like door handles, tail-lamp clusters, steering wheels, switch gear, suspension items, brakes and wheels, not to mention major assemblies such as engines, gearboxes and the internals of the various final-drive units, to be found from other sources if David Gerald TVR or a TVR dealer cannot immediately supply them. A new TVR owner can find hours of harmless amusement in prowling around his new car to see what 'hang-on' fittings he can recognize; if he cannot quite place something, he may be sure that the TVR Car Club, or David Gerald, has already done so!

The two major TVR-designed items, which are quite obviously not available from another manufacturer, are the bodyshells and the chassis-frames. The bad news here for the owners of early models is that when the factory sold its supply of jigs and moulds to David Gerald TVR, there was no trace of the 'trailing-link' frames or the body moulds to suit them and, worse, it seems that no drawings of this frame were ever made. Even so, Gerald Jinks has a rotten, rusty, trailing-link frame in pieces which could be copied. It is also understood that a mould for the early Grantura still exists, and that the TVR Car Club knows where to obtain supplies.

However, in 1993 I learned from Roger Cook, the TVR Car Club's Secretary, that a Grantura owner in East Anglia had actually constructed a jig when repairing the frame on his car, and that now, 30 years after production ceased, it should therefore be possible for newly-manufactured frames to be made once again.

The fact that in 1987 David Gerald TVR took over responsibility for maintaining the parts supply situation for all TVRs built in the 1962–79 period has transformed the possibilities of restoration, in particular for owners of Vixens, Ms, Taimars and 3000S Convertibles. Recently, when I challenged Gerald Jinks of David Gerald TVR with this question: "Could you build me a new 3000M, or 3000S from spare parts?", the succinct answer was: "Yes, of course – physically we could, but legally [because of VIN number regulations] we couldn't. All you need is patience – it will take a while – and money!"

The official TVR press release, dated March 23, 1987, tells it all: "In an unprecedented move TVR have released the official supply and manufacturing rights for all the pre-1980 spares to David Gerald TVR Sportscars Ltd ... As the sole officially appointed worldwide supplier, David Gerald will pursue the parts business ... All chassis and suspension jigs along with the vast number of body moulds and master bucks, for the fabrication and manufacture of these earlier parts are included in the deal."

Behind this historic deal was the fact that TVR were running out of space in Blackpool, needed to expand, and were also happy to release some capital in the process. It was no coincidence that the parts business was sold just as tooling expenditure on the new S model was reaching a climax.

Gerald Jinks himself takes up the story:

"David Field and I first met in about 1982 when we both went out and bought 3000Ms. At that juncture neither of us was in the motor trade – TVRs were just a hobby! Then we both went out and bought 3000Ss. Next I was persuaded to sell one car, and buy another, and soon what had been a hobby became a business.

"We started with a very small unit, then moved, then bought another unit – and in 1987 we bought all the spares from TVR for pre-1980 models. We decided to buy our place at Inkberrow, in Worcestershire, to house the TVR business; it had once been a grain storage barn!

"Before long we had filled right up, because people weren't only buying spares, but asking for service and repairs, so we installed a three-bay workshop at one end of the building. We had a new-car dealership with TVR at the end of the 1980s, but then I agreed with Peter Wheeler to release that, and we decided to concentrate on the older-car business – pre-Ms, Ms and early Tasmins.

"At that time, in any case, we were heavily into a motor racing programme, first with the 420SEAC that Steve Cole drove so successfully, and which Gerry Marshall also drove. Then the Tuscan Challenge came out, and we were fourth in the first series."

They were, in fact, very busy.

"Then David decided that he didn't want to stay – he wanted to go into Range Rovers – so we sold off the motorsport team, and we split off amicably. As you can see, the business now concentrates purely on the older-TVR business.

"Before we bought up all the stock from TVR, the factory was still doing a lot of servicing, but no restoration. The other new-TVR dealers still do some work on older cars, but most of the time they don't want to. They all buy parts from us though – we deal with the TVR trade *and* the TVR factory for some of their export inquiries."

Last time I called in at Inkberrow, a village not far east of Worcester, the stores were bursting at the seams with anything from a chassis-frame to a rubber bush, a body moulding to a tail-lamp cluster. Complete chassis-frames for cars as early as the Sixties-type Griffith are held in stock, and replacements can very quickly be built up, using the original factory jigs and fixtures – at least 50 chassis a year are delivered. There is full coverage from the early Sixties to the end of the M/Taimar/3000S period of the late Seventies. It is fair to say that DG would much prefer to supply completely new frames for repair or rebuilding purposes, as a frame which has been crashed badly, 'straightened' by less than expert hands, or which has suffered corrosion over many years, may never be the same again. In any case, as Jinks told me: "We can provide 99 per cent of everything for an M-Series car, straight away, then we could soon produce the rest.

The Ford 'Kent' engine has been built in many forms over the years, but from 1967 the 1,599cc unit was made available in crossflow form. In this Cortina 1600GT tune it was fitted to the Vixen range from 1967 to 1973, and to the 1600Ms built intermittently between 1972 and 1977. Over the years there were many minor development changes, but the basic form, including the downdraught compound dual-choke Weber carburettor and the tubular four-branch exhaust manifold, remained the same. This particular example was manufactured in 1970. (*Ford Motor Company*)

We've never refused a request yet. We've improved on the situation which existed before 1987 – though it was good – and we have actually managed to reduce costs in some cases.

"I can sell owners a Parts List, which has drawings and tells everything. That, plus the Handbook, tells everything for the M. TVR never did a Parts List for the Vixens, but we have had a Parts List built up for the 1960s-type Griffith."

How easy was it to trace where all the parts were originally sourced?

"TVR didn't supply us with any drawings, but they were very helpful. We were provided with all the build sheets – in some cases there had never been any drawings anyway! In some cases, when we wanted to have parts remade we had to approach the original manufacturers. In some cases the companies had already gone into liquidation – rear hub castings for Ms are a good example. We are very rarely stuck, and when we settle one problem, another one appears."

Does it help to be placed in Worcestershire, close to the

More TVRs have been built with the British Ford 3-litre 'Essex' V6 engine than any other type. It was usually fitted in the form illustrated, with a single downdraught two-choke Weber carburettor, and gave about 138bhp, but a select few of these units were extensively reworked by Broadspeed to produce 230bhp with the aid of turbocharging. Some details of TVR-installed engines may differ from this unit, which was as normally used in Ford Zodiacs and Granadas. (*Ford Motor Company*)

industrial Midlands?:

"Yes, definitely. *And* we're also in the middle of the country, so that we're relatively easy for everyone to find."

To browse around the stores and the workshops of David Gerald TVR is to realize that no-one should have an excuse for putting a TVR off the road for a lack of parts. New parts, rebuilt parts (engines and gearboxes) for instance, or in some cases reclaimed parts are available in profusion.

Although no VW suspension parts are held in stock, David Gerald will help identify which parts are absolutely the same as those used in the VW Beetle, for which the parts supply system is extremely comprehensive. There is no difficulty in finding suspension parts for the coil-spring chassis, of whatever type, though it is pointed out that nowadays, of course, no British manufacturer is building wire-spoke wheels for new cars, and that although new wheels *can* still be found by ordering through the various specialists, the day will inevitably come when that aspect of the car becomes a problem.

The supply situation in regard to bodyshells, or removable sections such as doors and the bonnet, is even more encouraging. Because all TVRs had glassfibre bodies and, like other cars, seemed to suffer from 'corner' damage in shunts, a whole series of part-bodyshell sections were made available from time to time and this service has continued. Complete bodies can be supplied covering the period from the first of the Grantura Mark IIs to the present day, and almost any section or part-section is also available, given a little notice. In the case of body styles, incidentally, be sure that the car you are considering buying is original in every respect. Due to the gradual way in which Grantura, Vixen and Tuscan styling evolved during the Sixties, it is quite likely that a car may have been updated or modified by the addition of a different bonnet moulding, later tail-lamp clusters, the substitution of an instrument layout from another model or the use of entirely different seats from another TVR. Most people like to see complete originality in their classic cars; in the case of older and once-neglected TVRs it might be time-consuming – and expensive – to return them to that state.

Almost all the major mechanical components – engines, gearboxes and the like – were taken, unmodified, from larger manufacturers. There should be no lasting difficulty, for instance, in rebuilding an MG engine used in a Grantura, or even one of the older type of four-cylinder Ford units. Nor, for that matter, should it ultimately be impossible to rebuild an American Ford V8 engine and transmission, but the wait for essential spare parts may take much longer; there are concerns in the UK and, of course, in the United States, which specialize in these big units.

The supply of trim and electrical items is also fairly straightforward. TVR made their own wiring looms and new looms can be supplied for almost any obsolete model, given time; some looms, particularly those used in the more recently obsolete models, are usually held in stock.

Unless money is a serious limiting factor, there should be no reason why *any* TVR, no matter how old or in what sort of disreputable condition, should not be rebuilt to a thoroughly roadworthy condition, and once this has been done it should also be easy enough to make sure it is properly maintained, as the factory would recommend. Such chores are made easier by the fact that there is a great deal of factory-sourced or factory-approved literature available to cover the subject. In preparing this book, for instance, a whole range of Owners' Handbooks were sifted through which deal with TVRs of all ages, that for the original series-production Granturas – Marks I, II, IIA and III – having been reprinted by TVR Engineering at the request of the TVR Car Club.

This brings me neatly to the subject of the TVR Car Club itself. Every buyer of a specialist car, particularly if it is obsolete, should join the specialist car club connected with his chosen marque. The investment in annual subscriptions can pay off many times over by the special knowledge which is relayed to all members. Earlier in this chapter it was made clear that in the case of TVR it would be very difficult to remain in ignorance of the club, as so many factory employees, or directors, are also influential members of the TVR Car Club.

In the Nineties the TVR Car Club Ltd is a thriving, apparently ever-expanding organization, but in the Eighties it went through a series of financial traumas. In 1993 Roger Cook, Club Secretary, revealed just how far matters have progressed. Most important of all was the setting-up of a permanent HQ at Cook's office in Cheltenham:

TVR Car Club Ltd
(Roger Cook, Secretary)
21 Hawkeswood Road
Woodlands
Cheltenham
Gloucestershire GL51 5DT

Tel: 0242–222878

Like most one-make clubs, the TVR Car Club seeks to provide spares and service for the cars, a forum for sales and wants – of cars and components – and leans more towards the preservation of the cars in the marque. Its atmosphere is closer to that of a social club than a motorsport-orientated group.

The Club became a limited company in 1989, at which point Roger Cook, a self-confessed TVR fanatic, stepped in as the Secretary. Roger, by the way, is a typical one-make club enthusiast, owning three TVRs and having been an enthusiast for the marque for many years. At that point the membership was about 1,250, but within four years it had risen to 3,000, and at the time of writing it was expanding fast. In the early Nineties, many buyers of new TVRs – S customers in particular – are joining the Club. Roger Cook reckons that more than 40 per cent of members use their TVRs as 'toys', rather than as day-to-day working machines; at least one member is known to own six different cars!

Nor is club membership exclusively British. While more than half *are* from the UK, there are also members in Europe (including Germany, Sweden and Switzerland), Australia, Japan, New Zealand, Oman, Singapore – and more than 500 from the USA.

When this book was being prepared in 1993, the annual subscription to the Club was £25 (with a £5 joining fee for new members). Regular contact between the Club and its membership is assured by the publication of a very professional-looking magazine, *TVR Sprint*, which is published monthly.

Although the TVR factory has no stake in the Club, or any active part in its running, Chairman Peter Wheeler is also the President of the Club. This is really only a figurehead position; the principal contact with the factory comes through Roger Cook (from the Club), or from the factory through Brian Horner and Ben Samuelson:

"The relationship between the Club and the factory is as good as it has ever been," Roger Cook told me. "We are in touch the whole time, and we regularly get involved in publicity work, particularly if a dealer needs an older model to brighten up one of his displays. Basically, if we want anything *other* than money, TVR usually tries to provide it."

The largest TVR dealers – such as Barnet Motor Co (north of London), Team Central in Birmingham, Graypaul Motors, and Harrogate Horseless Carriages – all support the Club's activities, sometimes slipping Club publicity leaflets into the glove pockets of cars going to new customers.

As far as the Club is concerned, in the UK it is split into 29 Regions, with almost every one getting together at least once a month. A mass visit to the factory is set up every other year, and the National Weekend is held annually at a different location. Track days, at racing circuits, also give owners a chance to unleash their TVRs in controlled conditions. Technical seminars, a meet usually concentrating on one particular subject, help enormously for those restoring the cars.

A recent development which, if other clubs' experiences are to be trusted, is bound to be a great incentive to joining the TVR Car Club is that a special insurance scheme is now operating in the UK. For the more modern TVRs, which are described fully in Volume 2, this is clearly a very important service.

Anyone with a TVR of this period should be able to keep the car in good condition by being a member of the TVR Car Club, and by making contact with David Gerald TVR. Even so, there may be occasions when contact with the factory is required.

TVR, situated at the address below, is a small concern with an informal organization, so there is no need to ask for a particular person. In particular, don't expect a personal reply from Peter Wheeler – he's far too busy!

TVR Engineering Ltd
Bristol Avenue
Blackpool
Lancashire FY2 0JF

Telephone : 0253–56151
Fax : 0253–57105
Telex : 647519

APPENDIX A

TVR milestones and the family tree: 1947–80

To summarize the complex technical and commercial events which have taken place at TVR since inception, this Appendix should form a useful 'skeleton' around which the main story and the details can be filled in:

1947 Trevor Wilkinson formed TVR Engineering and built his first 'special'.
1949 First TVR sold – to Trevor Wilkinson's cousin.
1954 Austin A40-engined TVR Sports Saloons first sold.
1957 Six prototype TVR Coupes sold.
1958 Grantura Mark I put into series production. TVR Engineering dissolved. Layton Sports Cars Ltd formed to take its place.
1961 Beginning of the Aitchison-Hopton period of ownership. Layton Sports Cars dissolved and TVR Cars Ltd founded.
1962 Announcement of Mark III coil-spring chassis and short-lived competitions programme. Trevor Wilkinson left the firm in April. Bankruptcy of TVR Cars Ltd in the autumn. Grantura Engineering Ltd (a creditor of the bankrupt company and its main supplier) took up the manufacture of TVRs.
1963 First deliveries of Griffiths to the USA.
1964 Announcement of Manx-tail derivative of Grantura – 1800S.
1965 Trident prototype revealed, but lost to other hands during the year. Collapse of Grantura Engineering in August. Purchase of all assets by Arthur and Martin Lilley in November. Formation of TVR Engineering Ltd.
1966 First 'Lilley-management' TVRs assembled – 1800S models.
1967 First new-model developments – Vixen replacing Grantura 1800S; Tuscan V8 replacing obsolete Griffith 400. Tina prototypes shown at Earls Court motor show, but not progressed.
1968 First showing of wide-bodied Tuscan V8 SE. Body never used on any other TVR, but was effective forerunner of M-type style.
1969 Tuscan V6 introduced – halfway model between Vixen S2 (with 1.6-litre four-cylinder engine) and Tuscan V8 SE (with 4.7-litre V8 engine).
1970 Assembly facilities moved from original Hoo Hill factory to Bristol Avenue, Bispham, works – completed over the Christmas period.
1971 1300 and 2500 first sold, on basis of Vixen S3 chassis and style: replaced neither Tuscan nor Vixen S3.
1972 Introduction of M-Series chassis and related body style, all-new in detail, but still with coil-spring front and rear independent suspension. Pre-M-Series chassis (1962–72) immediately phased out, with no overlap. 2500M was original M-Series chassis, soon joined by 1600M and 3000M. Some 1300, 2500 and Vixen (S4) models built with M-Series chassis and old body style – this anomaly cleared up by May 1973
1973 1600M discontinued (but re-introduced in 1975).
1975 Serious disruption due to fire at Bristol Avenue on January 3, 1975. Production restarted in March/April. TVR Turbo (with Broadspeed-modified 3-litre Ford engine announced at Earls Court motor show. Reintroduction of 1600M model.
1976 Introduction of Taimar with hatchback.
1977 Last de-toxed 2500M built (for the USA) as engine taken out of production by Triumph. 3000M took its place.

1978 Introduction of TVR Convertible – first drophead series-production TVR, which instantly took over majority of TVR sales.

1979 Last M-Type/Taimar/Convertible models built prior to launch of all-new TVR in 1980.

1980 Launch, in January, of Tasmin – new chassis, new body style, new engine (German Ford, 2.8-litre V6 with fuel injection) – no carry-over of components, except in minor detail, from M-Type range. Two-seater coupe at first, joined in October (at NEC motor show) by two-seater convertible and 2+2 coupe.

Notes:
Tuscan V8 was mechanically similar to Griffith 400.
Vixen S4 had Vixen S3 mechanicals and bodyshell on 1600M rolling chassis.
1600M was out of production from 1973–75.
Turbocharged engines available from 1975–76 to end of 3000M/Taimar/Convertible production.
Between 1958 and 1980 there were four basically different chassis – the VW-suspension type, the Grantura Mark III-type, the M-Series and the Tasmin-type.

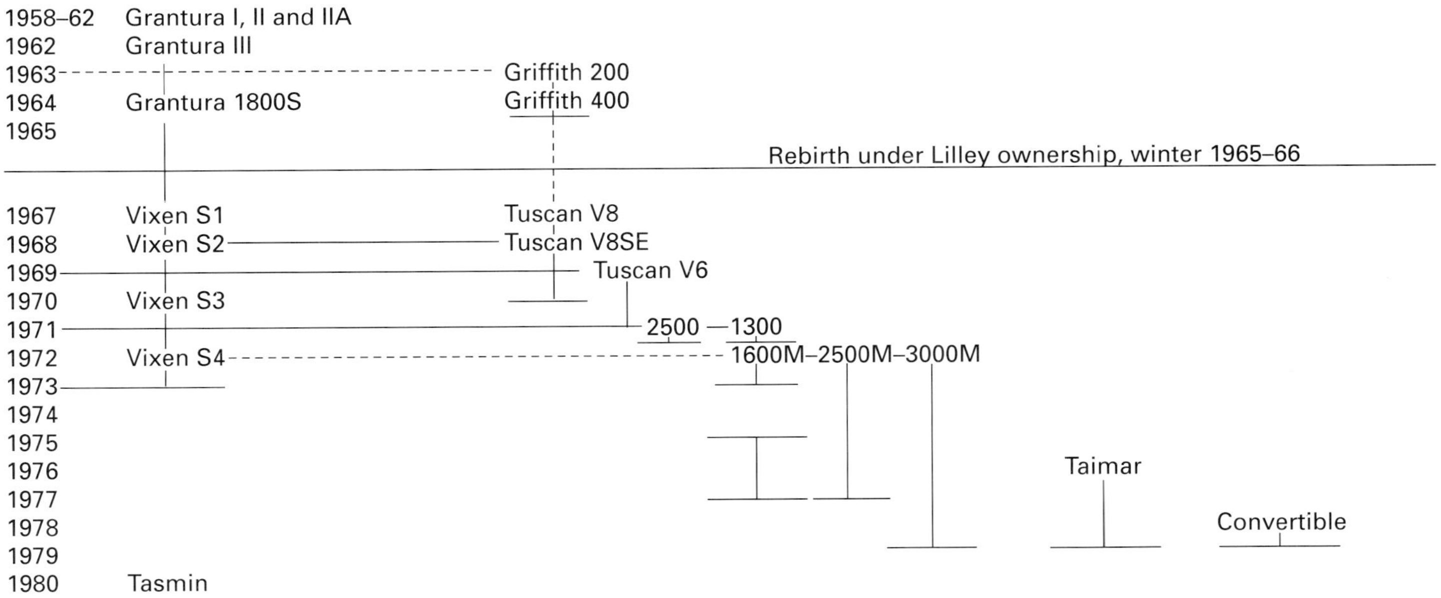

APPENDIX B

Technical specifications

Grantura Mark I – produced 1958 to 1960
Engine: Choice of Coventry Climax, Ford or BMC. Principal engine: Coventry Climax Type FWE, 4-cyl, 76.2 x 66.7mm, 1,216cc, CR 10:1, 2 SU carbs, single-overhead-camshaft valve gear. 83bhp (nett) at 6,000rpm. Maximum torque 75lb.ft at 4,800rpm.
Alternative engines: Ford (side-valve) 100E unit of 1,172cc, 35bhp at 4,600rpm, with Shorrock supercharging 56bhp at 4,600rpm; Ford (overhead-valve) 105E unit of 997cc, 39bhp (nett) at 5,000rpm; BMC (MGA-type overhead-valve) unit of 1,489cc, 72bhp (nett) at 5,500rpm, or Coventry Climax (overhead-camshaft) Type FWA of 1,098cc.
Transmission: Axle ratio 4.7:1 with Coventry Climax engine, other ratios between 4.1:1 and 5.1:1 also available depending on engine option chosen, hypoid-bevel gears. Four-speed BMC B-Series gearbox with Coventry Climax and MGA engines, three-speed Ford gearbox with 100E side-valve engine, four-speed Ford gearbox with 105E overhead-valve engine. Overall gear ratios with principal Coventry Climax engine 4.70, 6.46, 10.41, 17.11, reverse 22.37:1, with synchromesh on top, third and second gears. Close-ratio gears also available. 15.32mph/1,000rpm in top gear.
Suspension and brakes: Ifs, transverse torsion-bars, twin trailing-arms, telescopic hydraulic dampers. Irs, transverse torsion-bars, twin trailing-arms, radius-arms, telescopic hydraulic dampers. Worm-and-sector steering. 11 x 2.25in front and rear brakes; 11 x 1.75in on later cars. 5.00-15in or 5.60-15in tyres on 4.5in rims. Centre-lock wire-spoke wheels.
Dimensions: Wheelbase 7ft; front track 4ft 3in; rear track 4ft 4in. Length 11ft 6in; width 5ft 4in; height 4ft. Fuel tank 8.75gall. Unladen weight 1,455lb with Coventry Climax engine, 1,570lb with MGA engine.
Basic price: Kits £950 (with Coventry Climax engine), £800 (with MGA engine), £695 (with Ford 105E engine), £660 (with Ford 100E engine).

Grantura Mark II – produced 1960 and 1961
Specification as for Grantura mark I except for:
Engine: Choice of BMC, Coventry Climax or Ford. Principal engine: BMC, MGA 1600 unit, 4-cyl, 75.39 x 88.9mm, 1,588cc, CR 8.3:1, 2SU carbs, pushrod-operated overhead valve gear. 79.5bhp (nett) at 5,600rpm. Maximum torque 87lb.ft at 3,800rpm. Alternative engines: Ford (overhead-valve) 105E unit, Coventry Climax (overhead-camshaft) FWE unit; a few MGA 1500 engines of 1,489cc, 72bhp (nett) at 5,500rpm, were also fitted.
Transmission: Axle ratio and gearbox depending on engine chosen. With MGA 1600 unit, axle ratio 4.3:1. Overall gear ratios 4.3, 5.908, 9.52, 15.652, reverse 20.468:1, with synchromesh on top, third and second gears, 17mph/1,000rpm in top gear. Optional axle ratios 4.1, 4.55, 4.875, 5.1:1. Optional close-ratio gears, or optional ZF four-speed gearbox.
Suspension and brakes: 11 x 1.75in drum brakes, front and rear. 5.60-15in tyres. Rack-and-pinion steering option on Coventry Climax-engined cars.
Basic price: Kits, £880 (with MGA engine), £1,045 (with Coventry Climax engine), £795 (with Ford 109E engine).

Grantura Mark IIA – produced 1961 and 1962
Specification as for Grantura Mark II except for:
Engine: Most cars built with MGA 1600 Mark II engine, 4-cyl, 76.2 x 88.9mm, 1,622cc, CR 8.9:1, 2 SU carbs, pushrod-operated overhead valve gear, 86bhp (nett) at 5,500rpm. Maximum torque 97lb.ft at 4,000rpm. Early cars available with MGA 1,588cc engine, plus previous Ford and Coventry Climax engine options. Further alternative engine, Ford (overhead-valve) 109E of 1,340cc, 54bhp (nett) at 4,900rpm.
Suspension and brakes: 11.25in front discs, 11 x 1.75in rear drums.

Grantura Mark III – produced 1962 to 1964
Engine: MGA 1600 Mark II engine to September 1963, MGB

1800 engine thereafter. Ford and Coventry Climax engine options listed, but very few cars built with those units.
MGA 1600 Mark II engine: 4-cyl, 76.2 x 88.9mm, 1,622cc, CR 8.9:1, 2 SU carbs, 86bhp (nett) at 5,400rpm. Maximum torque 97lb.ft at 4,000rpm.
MGB 1800 engine: 4-cyl, 80.26 x 88.9mm, 1,798cc, CR 8.8:1. 2 SU carbs, 95bhp (nett) at 5,400rpm. Maximum torque 110lb.ft at 3,000rpm.
Transmission: Axle ratios to choice, 4.1, 4.55, 4.875, 5.1:1. Normally 4.55:1 with 1,622cc engine and overall ratios 4.55, 6.25, 10.07, 16.56, reverse 21.66:1, 16.1mph/1,000rpm. Normally 4.1:1 with 1,798cc engine and overall ratios 4.1, 5.633, 9.077, 14.924, reverse 19.516:1. 17.8mph/1,000rpm in top gear. Synchromesh on top, third and second gears. Optional overdrive on top and third gears: overall overdrive top gear ratio with 4.1 axle, 3.30:1, 22.1mph/1,000rpm in overdrive top gear.
Suspension and brakes: Ifs, wishbones, coil spring/telescopic damper units and anti-roll bar. Irs, wishbones and coil spring/telescopic damper units. Rack-and-pinion steering. 10.75in front disc brakes, 9 x 1.75in rear drums. Centre-lock wire-spoke wheels or steel-disc wheels with 4.5in rims. 5.60-15in tyres.
Dimensions: Wheelbase 7ft 1.5in; front track 4ft 3in; rear track 4ft 4in. Length 11ft 6in; width 5ft 4in; height 4ft. Fuel tank 10gall. Unladen weight 1,625lb with MGA or MGB engine.
Basic price: Kits, £862 (with MGA 1600 Mark II engine), £872 (with MGB 1800 engine).

Grantura 1800S – produced 1964 to 1966, and 'Mk IV' 1800S – produced 1966 to 1967

Specification as for Grantura Mark III except for:
Engine: BMC MGB 1800 engine standardized with no options.
Transmission: Axle ratio 4.3:1 or 3.9:1, to choice. With 4.3:1 ratio, overall gear ratios 4.3, 5.908, 9.52, 15.652, reverse 20.468:1; optional overdrive on top and third gears, overall overdrive top ratio 3.45:1. 17mph/1,000rpm in direct top gear, 21.2mph/1,000rpm in overdrive top gear. With 3.91:1 ratio, overall gear ratios 3.91, 5.37, 8.66, 14.21, reverse 18.59:1; optional overdrive on top and third gears, overall overdrive top ratio 3.14:1. 18.7mph/1,000rpm in direct top gear, 23.3mph/1,000rpm in overdrive top gear.
Dimensions: Front track 4ft 4.5in; rear track 4ft 5.5in. Fuel tank (Mark IVS) 15gall. Unladen weight (Mark IVS) 1,790lb.
Basic price: £872 at first, £927 from February 1965, £981 from June 1965, £988 from revival of production in spring 1966.

Vixen S1 – produced 1967 and 1968

Engine: Ford Cortina GT 1600 engine, 4-cyl, 80.98 x 77.62mm, 1,599cc, CR 9:1, two-choke Weber carb, 88bhp (nett) at 5,400rpm. Maximum torque 96lb.ft at 3,600rpm.**
Transmission: Ford all-synchromesh gearbox with axle ratio of 3.89:1. Overall gear ratios 3.89, 5.45, 7.84, 11.59, reverse 12.91:1. 18.8mph/1,000rpm in top gear.
** A few cars were fitted with MGB 1,798cc engine and BMC gearbox, left over from Grantura Mark IVS production.
Suspension and brakes: Ifs, wishbones, coil spring/telescopic damper units and anti-roll bar. Irs, wishbones and coil spring/telescopic damper units. Rack-and-pinion steering. 10.75in front disc brakes, 9 x 1.75in rear drums. Steel-disc or centre-lock wire-spoke wheels with 4.5in rims. 165-15in tyres.
Dimensions: Wheelbase 7ft 1.5in; front track 4ft 4.5in; rear track 4ft 5.5in. Length 11ft 6in; width 5ft 4in; height 4ft. Fuel tank 15gall. Unladen weight 1,735lb.
Basic price: £988 at first, £1,095 from April 1967.

Vixen S2 – produced 1968 to 1970

Specification as for Vixen S1 except for:
Engine: MGB engine option not available.
Brakes: With servo-assistance from 1969.
Dimensions: Wheelbase 7ft 6in; front track 4ft 5in; rear track 4ft 6in. Length 12ft 1in; width 5ft 4in; height 4ft. Unladen weight 1,680lb.
Basic price: £1,095 at first, £1,150 from January 1969, £1,242 from May 1970.

Vixen S3 – produced 1970 to 1972

Specification as for Vixen S2 except for:
Engine: Ford power output now quoted as 86bhp (DIN) at 5,500rpm.
Maximum torque 92lb.ft (DIN) at 4,000rpm.

Suspension and brakes: Wheels, bolt-on cast-alloy with 5.5in rims.
Dimensions: Unladen weight 1,625lb.
Basic price: £1,245 at first, £1,295 from April 1971, £1,310 from July 1971.

1300 – produced 1971 and 1972
Specification as for Vixen S2 except for:
Engine: BL Triumph Spitfire unit, 4-cyl, 73.7 x 76mm, 1,296cc, CR 9:1, 2 SU carbs, 63bhp (DIN) at 6,000rpm. Maximum torque 69lb.ft at 3,500rpm.
Transmission: BL Triumph Spitfire all-synchromesh gearbox. Axle ratio 3.89:1. Overall gear ratios 3.89, 5.41, 8.41, 13.65, reverse 15.51:1. 17.8mph/1,000rpm in top gear.
Dimensions: Unladen weight 1,625lb.
Basic price: £1,245 at first, £1,355 from April 1972, £1,422 from July 1972.
Note: The last six 1300s were fitted with M-Series chassis – see below.

All the above TVRs had four-cylinder engines. The following cars fitted with the 1962–72 coil-spring chassis-frame were also produced.

Griffith 200 and Griffith 400 V8 – produced 1963 to 1965
Specification as for Grantura Mark III and 1800S except for:
Engine: USA Ford V8 units in two states of tune. 195bhp unit: V8-cyl, 101.6 x 72.9mm, 4,727cc, CR 9:1, four-choke downdraught Ford carb, 195bhp (gross) at 4,400rpm. Maximum torque 282lb.ft at 2,400rpm. 271bhp unit: CR 11.0:1, 271bhp (gross) at 6,500rpm. Maximum torque 314lb.ft at 3,400rpm.
Transmission: Axle ratio 3.07:1. Overall gear ratios 3.07, 4.33, 5.46, 7.25, reverse 7.25:1. 25mph/1,000rpm in top gear. Synchromesh on all forward gears.
Suspension and brakes: Extra spring/damper units on rear suspension. Servo-assisted brakes. 185-15in tyres on 5in wheels. All cars built with centre-lock wire-spoke wheels.
Dimensions: Fuel tank 17gall. Unladen weight 1,905lb.
Basic price: (Only sold in UK during 1965) £1,342 at first, £1,400 from June 1965.

Tuscan V8 – produced in 1967
Specification as for Grantura Mark III except for:
Engine: USA Ford V8 units as for Griffith 200/400 models.
Transmission: Ford gearbox as for Griffith 200/400 models.
Suspension, dimensions and weight: As for Griffith 400 model, except fuel tank 15gall.
Basic price: £1,599, or £1,922 as SE model with 271bhp (gross) engine.

Tuscan V8 SE – long-wheelbase – produced 1967 and 1968
Specification as for Griffith 400 and Tuscan V8 models except for:
Suspension and brakes: Wire wheels with 6in rim width.
Dimensions: Wheelbase 7ft 6in. Length 12ft 1in. Unladen weight 2,240lb.
Basic price: £1,922.

Tuscan V8 SE – long-wheelbase/wide-body – produced 1968 to 1970
Specification as for Griffith 400 and Tuscan V8 models except for:
Suspension and brakes: Wire wheels with 6in rim width.
Dimensions: Wheelbase 7ft 6in. Length 12ft 2in; width 5ft 8in.
Basic price: Not quoted – only three cars sold in UK.

Tuscan V6 – produced 1969 to 1971
Specification as for Vixen S2/S3 except for:
Engine: British Ford unit, V6-cyl, 93.66 x 72.4mm, 2,994cc, CR 8.9:1, two-choke Weber carb, 128bhp (DIN) at 4,750rpm. Maximum torque 173lb.ft at 3,000rpm.
Transmission: Ford all-synchromesh gearbox. Axle ratio (early cars) 3.31:1. Overall gear ratios 3.31, 4.67, 7.33, 10.47, reverse 11.08:1. Optional overdrive on top and third gears; overall ratio in overdrive top 2.72:1. 22mph/1,000rpm in direct top gear, 26.9mph/1,000rpm in overdrive top gear. Axle ratio (later cars) 3.54:1. Overall gear ratios 3.54, 5.00, 7.84, 11.20, reverse 11.84:1. 20mph/1,000rpm in top gear.
Dimensions: Unladen weight 2,000lb.
Basic price: £1,492 at first, £1,558 from May 1970, £1,595 from April 1971.

2500 – produced 1971 and 1972
Specification as for Vixen S2/S3 except for:
Engine: Triumph TR6 (federal-tune) engine, 6-cyl, 74.7 x 95mm, 2,498cc, CR 8.5:1, 2 Zenith-Stromberg carbs, 106bhp (nett) at 4,900rpm. Maximum torque 133lb.ft at 3,000rpm.
Transmission: Triumph rear axle and gearbox. Axle ratio 3.45:1. All-synchromesh gearbox with optional overdrive on top and third gears. Overall gear ratios (o/d 2.83), 3.45, 4.59, 6.94, 10.83, reverse 11.11:1. 21.1mph/1,000rpm in direct top gear, 25.7mph/1,000rpm in overdrive top gear.
Dimensions: Unladen weight 1,960lb.
Basic price: £1,500 at first, £1,430 from October 1971, £1,560 from May 1972, £1,638 from August 1972.
Note: The last series of 2500s, built in the spring and summer of 1972, had M-Type chassis – see below.

From the spring of 1972, a new basic chassis-frame, known as the M-Type chassis, was standardized for all new TVRs. The M-Type range consisted of 2500M, 1600M, 3000M, Taimar and Convertible derivatives.

2500M – produced 1972 to 1977
Engine: Triumph TR6 (federal-tune) unit, 6-cyl, 74.7 x 95mm, 2,498cc, CR 7.75:1 (from 1974, CR 7.5:1), 2 Zenith-Stromberg carbs, 106bhp (nett) at 4,900rpm. Maximum torque 133lb.ft at 3,000rpm.
Transmission: Triumph rear axle and gearbox. Axle ratio 3.45:1. All-synchromesh gearbox with optional overdrive on top and third gears. Overall gear ratios (o/d 2.83), 3.45, 4.59, 6.94, 10.83, reverse 11.11:1. 21.1mph/1,000rpm in direct top gear, 25.7mph/1,000rpm in overdrive top gear.
Suspension and brakes: Ifs, wishbones coil spring/telescopic damper units and anti-roll bar. Irs, wishbones, coil spring/telescopic damper units. Rack-and-pinion steering. 10.9in front disc brakes, 9 x 1.75in rear drum brakes, with vacuum servo assistance. Cast-alloy bolt-on disc wheels with 5.5in rims and 165-15in tyres; from autumn 1973, 6in wheel rims and 185HR-14in tyres.
Dimensions: Wheelbase 7ft 6in; front track 4ft 5.75in; rear track 4ft 5.75in. Length 12ft 10in; width 5ft 4in; height 4ft. Fuel tank 12gall. Unladen weight 2,240lb.
Basic price: £1,560 at first, £1,695 when into production in spring 1972, £1,779 from August 1972. Discontinued in UK in May 1973.

Vixen S4 – produced in 1972 and 1973
Specification as for 2500M except for:
Engine: Ford 1600GT overhead-valve unit, 4-cyl, 80.98 x 77.62mm, 1,599cc, CR 9:1, two-choke Weber carb, 86bhp (DIN) at 5,500rpm. Maximum torque 92lb.ft at 4,000rpm.
Transmission: Unladen weight 1,970lb.
Basic price: £1,560 at first, £1,638 in August 1972 and 1973, £2,200 when reintroduced in 1975, rising rapidly to £2,975 by autumn 1976.

3000M – produced 1972 to 1979
Specification as for 2500M except for:
Engine: British Ford unit, V6-cyl, 93.66 x 72.4mm, 2,994cc, CR 8.9:1, two-choke Weber carb, 138bhp (DIN) at 5,000rpm. Maximum torque 174lb.ft at 3,000rpm.
Transmission: Ford all-synchromesh gearbox. Axle ratio 3.45:1. Overall gear ratios 3.45, 4.86, 10.90, reverse 11.56:1. 21.6mph/1,000rpm in top gear. Overdrive not available until autumn 1975.
Dimensions: Unladen weight 2,240lb.
Basic price: £1,795 in mid-1972, rising to £6,417 when discontinued in 1980.

Taimar – produced 1976 to 1979
Specification as for 3000M except for hatchback body and:
Engine: 142bhp (DIN) at 5,000rpm.
Dimensions: Unladen weight 2,260lb.
Basic price: £4,260 in autumn 1976, rising to £7,211 when discontinued in 1980.

Convertible – produced 1978 and 1979
Specification as for 3000M except for convertible body and:
Engine: 142bhp (DIN) at 5,000rpm.
Dimensions: Height 3ft 9in. Unladen weight 2,420lb.
Basic price: £5,462 in mid-1978, rising to £7,007 when discontinued in 1980.

Note: From 1976 to 1980, a turbocharged engine option was available on 3000M, Taimar and Convertible models. Specification differences were as follows:
Engine: CR 8:1, two-choke Weber carburettor with Broadspeed/Holset or AiResearch turbocharger installation. 230bhp (DIN) at 5,500rpm. Maximum torque 273lb.ft at 3,500rpm.
Transmission: Axle ratio 3.31:1. Overall gear ratios 3.31, 4.67, 6.45,10.46, reverse 11.09:1. 21.7mph/1,000rpm in top gear. 195/70VR-14in tyres.
Basic price: Extra charge for installation £2,490 in spring 1976, rising to £3,210 by 1980.

At the end of 1979, the M-Series chassis was discontinued and TVR introduced their first all-new chassis and bodyshell since the late Fifties. This was the Tasmin, announced in two-seater fixed-head form in January 1980 and later available in other derivatives. (See Volume 2.)

Gerald Jinks (right) and David Field secured the future of old-TVR restoration in 1987 when their business, David Gerald TVR, took over the rights to parts supply and manufacture for all pre-Tasmin models. (*Gerald Jinks*)

APPENDIX C

Chassis number sequences

All the staff at TVR gave every possible assistance in compiling the following table of Chassis Number sequences. However, it should be explained that existing records held by TVR only originate with the rebirth of the company in the winter of 1965–66 and that earlier information has been supplied from the memory of that remarkable TVR staff man, Stan Kilcoyne.

Once TVR was started up again by the Lilley family, an entirely new sequence of Chassis Numbers was adopted. With the exception of the Ford V8-engined cars (Griffith and Tuscan V8 types), which had their own separate sequences, all TVRs built from 1966 to 1979 took up the appropriate number issued when that particular car was built. The sequence started from ...001 in February 1966 and ended at ...4970 in December 1979, with a single hiccup: the last Vixen Series 1 was VX233, while the first Vixen Series 2 which followed it was VX1234/F. In other words there was a jump of 1,000 numbers at that point, in October 1968.

What this means is that not all cars built between, say, LVX 1283/6 and LVX 2134/6 were Tuscan V6 models, but that all Tuscan V6s were contained inside the spread of these numbers. In case any TVR owner suspects that the car he buys is not what it seems, he should by happy to know that the record books have been preserved at Blackpool, and it was these which were consulted in preparing the table:

Model	Years built	Engine	Chassis Numbers
Grantura Mk I	1958–60	MGA/MGB	7B
Grantura Mk II	1960–61	Coventry Climax	7C – 001 onwards
Grantura Mk IIA	1961–62	Ford	7F
Grantura Mk III	1962–64	MGA/MGB	
Grantura 1800S	1964–66	MGB	18–001 onwards
Griffith 200	1963–64		200–001 onwards
Griffith 400	1964–65	Ford V8	400–001 onwards
Griffith 400	1965		200/GB/5001 onwards

Following rebirth under the management of the Lilley family:

Grantura 1800S	1966	MGB	18/001–18/044
Grantura 'Mk IV' 1800S	1966–67	MGB	18/027–18/116
Vixen S1	1967–68	Ford 1600	VX117–VX233
Vixen S2	1968–70	Ford 1600	LVX1234/F –LVX1736/4
Vixen S3	1970–72	Ford 1600	LVX1737/4– LVX2239/4
Vixen S4	1972–73	Ford 1600	2271/4–2597/4
Griffith	1966–67	Ford V8	200–001 to 200–010
Tuscan V8	1967	Ford V8	200–011 to 015, 017 to 020, 022 to 040
Tuscan V8 lwb	1967–68	Ford V8	LWB–001 to LWB–024
Tuscan V6	1969–71	Ford V6 3.0	LVX1283/6 –LVX2134/6
Tuscan V8 (wide-body)	1968–70	Ford V8	MAL001–MAL021
2500	1970–72	Triumph 2500	1745/6T–2703/T
1300	1971–73	Triumph 1300	1992/S–2557/4S
2500M	1972–77	Triumph 2500	2090T, then 2240TM–4094TM
1600M	1972–73	Ford 1600	2288FM–2623FM, then
and	1975–77		3384FM–3938FM
3000M	1972–79	Ford V6 3.0	2410FM–4940FM
Taimar	1976–79	Ford V6 3.0	3838FM–4966FM
Convertible	1978–79	Ford V6 3.0	4286FM–4968FM

Notes:

1. In the case of the early Granturas, the first letter in the sequence pinpoints the engine – B for BMC (MG) type, C for Coventry Climax and F for Ford, of whatever type.

2. MGB-engined cars are pinpointed by the prefix 18 in the Chassis Number sequence.

3. Series 1 Vixens are identified by the prefix VX. When the Vixen Series 2 was introduced, the wheelbase was increased, so it was logical that the prefix should be LVX. The Tuscan V6 and the 2500 model both used the same basic chassis, and the V6 also adopted the LVX prefix.

4. Progressively, from the introduction of the Vixen Series 2, the chassis prefixes were phased out and replaced by a series of suffixes:
Vixen 2, 3 and 4 models carried suffix .../4 (early cars .../F)
Tuscan V6 models carried suffix .../6
2500 models carried suffix .../6T, later reduced to .../T
1300 models carried suffix .../S (or .../4S)

5. All cars based on the M-Series chassis (except the 1971 prototype 2500M) used numbers ending with ...M, as follows:
2500M models carried suffix ...TM
1600M models carried suffix ...FM
3000M models carried suffix ...FM
Taimar models carried suffix ...FM

6. A search of the chassis records confirms that almost all Convertibles had even-number Chassis Numbers. However, the occasional 3000M or Taimar also slipped into this sequence. In general, though, from early 1978, the Convertible took over all even numbers from 4286FM. Because of the imbalance caused by the above arrangement, the last odd number to be issued was 4699FM, and this was, in fact, a Convertible!

7. There was a clear cut-off point between pre-M and M-Series chassis. The last pre-M chassis was 2239 (a Series 3 Vixen), and the first M-Series chassis was 2240TM (a 2500M); there was no overlap.

8. Identification of the Turbos – turbocharged engines supplied on 3000M, Taimar and Convertible bases – is considered interesting enough for all the cars to be listed individually, as shown below.

9. Following the disastrous fire at the Bristol Avenue works, on January 3, 1975, there was a considerable hiatus before production could recommence. The first 'post-fire' car to be completed – a 3000M – carried Chassis Number 3380FM.

10. The move to Bristol Avenue took place over Christmas 1970. The first car to be completed at the new premises carried Chassis Number LVX 1806/6 and was a Tuscan V6.

11. After the introduction of the M-Series chassis in April 1972, some Vixen, 1300 and 2500 models were built with M-Series chassis and old-style bodies. Details were as follows:

Vixen S4 – 23 cars	from 2271/4 to 2597/4
1300 – 6 cars,	culminating in 2557/4S
2500 – 96 cars	from 2258T to 2703T

How to keep the troops amused on a wet day! TV and radio personality – and TVR enthusiast – Stuart Hall holds centre stage at a TVR club gathering, while proprietor Martin Lilley is happy that he does not have to do the same. (*TVR*)

The exclusive Turbo model

Between September 1975 and December 1979, a total of 63 cars were built with turbocharged 3-litre engines. Their Chassis Numbers and body styles were as follows:

3463	3000M	4125	Taimar	4471	Taimar
3569	3000M	4170	3000M	4488	Convertible
3587	3000M	4197	Taimar	4537	Taimar
3597	3000M	4207	Taimar	4550	Convertible
3604	3000M	4243	Taimar	4565	Taimar
3606	3000M	4244	3000M	4566	3000M
3620	3000M	4254	3000M	4567	Taimar
3789	3000M	4279	Taimar	4571	Taimar
3815	3000M	4298	3000M	4608	3000M
3837	Taimar	4311	Taimar	4647	Taimar
3840	Taimar	4313	Taimar	4649	Taimar
3841	3000M	4325	Taimar	4661	Taimar
3844	Taimar	4327	Taimar	4673	Taimar
3850	Taimar	4332	3000M	4697	Convertible
3853	Taimar	4338	3000M	4704	Convertible
3957	Taimar	4354	Convertible	4746	Convertible
3961	Taimar	4362	Convertible	4768	Convertible
3965	Taimar	4404	Convertible	4800	3000M
3999	Taimar	4422	Convertible	4808	Convertible
4015	Taimar	4423	Taimar	4938	3000M
4105	Taimar	4468	Convertible	4970	Convertible

Of the 63 cars, 20 were 3000Ms, 30 were Taimars and 13 were Convertibles.

Chassis Numbers 3850, 4298, 4423 and 4647 were left-hand drive, all others being right-hand drive.

There were four 'SE' or Special Equipment Turbos – Chassis Numbers 4327, 4537, 4649 and 4768.

APPENDIX D

Production and deliveries

Two factors make it difficult to present a detailed and entirely accurate summary of TVR production from 1958 when series production of Granturas began. One is that the tempestuous commercial history which was normal at TVR throughout the 1958-65 period means that the present management does not have any of the old records. The other is that TVR Engineering Ltd has latterly detailed its chassis records not by the date at which a certain car was completed and delivered, but by the date when the production works *order* was issued to the workshops. Therefore, full and accurate production totals have been given for each different and identifiable model and – wherever possible – this is broken down into the number of cars *ordered into production* in any particular calendar year.

Annual production figures

TVR became members of the Society of Motor Manufacturers and Traders in 1967, and the SMM&T have been able to provide the following figures for the calendar years 1970 to 1980:

Year	Total production	Home market	Export market
1970	284	190	94
1971	360	181	179
1972	388	220	168
1973	388	101	287
1974	421	36	385
1975	132*	91	41
1976	333	186	147
1977	366	213	153
1978	310	164	146
1979	308	168	140
1980	144**	133	11

* As a result of the fire of January 1975.
** As a result of the major model changeover to Tasmin.

Total production of each model

For TVR models before the end of 1965 the totals quoted are approximations. Thereafter, it has been possible to provide exact figures:

Model	Total production	Comments
Grantura Mk I	100	
Grantura Mk II Grantura Mk IIA	} 400 approx.	
Grantura Mk III	90 approx.	
Grantura 1800S	90 approx.	
Griffith	300 approx.	(All but approx. 20 exported)

From 1966, under the control of the Lilley family:

Grantura 1800S	38	Including 7 right-hand-drive cars, all built in 1966
Grantura 'Mk IV' 1800S	78	40 LHD, 38 RHD, built in 1966 and 1967
Griffith	10	4 RHD, 6 LHD, built in 1966 and 1967
Vixen S1	117	12 with MGB engine, 105 with Ford 1600GT engine
Vixen S2	438	47 delivered in 1968 209 delivered in 1969 182 delivered in 1970
Vixen S3	168	50 delivered in 1970 97 delivered in 1971 21 delivered in 1972
Vixen S4	23	22 ordered in 1972 1 ordered in 1973
Tuscan V8	28	4 RHD, 24 LHD, all built in 1967
Tuscan V8 lwb	24	12 each, RHD and LHD, 1967 and 1968
Tuscan V8 (wide-body)	21	All but two with LHD, built 1968 to 1970
Tuscan V6	101	32 delivered in 1969 50 delivered in 1970 19 delivered in 1971
2500	385	8 ordered in 1970 237 ordered in 1971 44 ordered in 1972 – the above 289 cars having old-type (Vixen-style) chassis. Plus a further 96 cars with M-type chassis and old-style bodies: 95 ordered in 1972 1 ordered in 1973

Model	Total production	Comments

All annual figures quoted below are for cars *annually ordered into production:*

1300	15	9 with old-style (Vixen) chassis 6 with M-Type chassis, all but last car with old-style bodies
2500M	947	1 in 1971 112 in 1972 263 in 1973 372 in 1974 20 in 1975 103 in 1976 76 in 1977
1600M	148	57 in 1972 11 in 1973 - in 1974 50 in 1975 29 in 1976 1 in 1977

3000M	654	28 in 1972
		120 in 1973
		49 in 1974
		84 in 1975
		161 in 1976
		132 in 1977
		42 in 1978
		38 in 1979
Taimar	395	15 in 1976
		155 in 1977
		125 in 1978
		100 in 1979
Convertible	258	129 in 1978
		129 in 1979
3000M Turbo	20	1 in 1975
		9 in 1976
		3 in 1977
		5 in 1978
		2 in 1979
Taimar Turbo	30	5 in 1976
		7 in 1977
		10 in 1978
		8 in 1979
Convertible Turbo	13	7 in 1978
		6 in 1979

APPENDIX E

Performance figures for all models

Assembling accurate and impartial sets of performance figures for TVRs has been fascinating, not only for what they reveal, but because there are so many different models to compare with each other over the years.

However, in order that performance data for as many models as possible could be included, the rule which was rigidly applied to my previous *Collector's Guides* has been relaxed. Even though they managed to test many TVRs between 1961 and 1980, *Autocar* never got their hands on an MG-engined Grantura, or either of the Tuscans (V6 or V8). It would have been wrong not to include some perfectly good statistics for these interesting cars, so *Motor* was consulted, who did test these cars.

It should, however, be emphasized that *Motor*'s testing methods differed slightly from those of *Autocar* until the end of the Sixties, when they acquired a fifth wheel, so there may be slight inconsistencies in the figures quoted.

During the Seventies, the authoritative American magazine *Road & Track* tested several TVRs, and their figures for the 2500, 2500M, Convertible and Tuscan V8 are also included. It should be pointed out, however, that a happy quirk in USA 'federal' regulations at the end of the Seventies allowed TVR to send in supplies of their 3-litre Ford V6-engined cars with modifications done locally by Olsen Laboratories of California.

It is very disappointing not to be able to include figures for the most powerful Griffith model, as it does not appear to have been tested by any of the magazines whose methods are trusted implicitly by me. However, *Motor*'s Tuscan V8 of 1967 was effectively the same car as the Griffith 400 except that it was somewhat heavier, and was therefore not quite as shatteringly fast in acceleration through the gears.

Both *Motor*'s Tuscans were given 'estimated' maximum speeds, the reason (in each case) being that the magazine had not journeyed to the Continent to find a limit-free motor road on which to establish the true top speed. In the case of the V6, the car would have been over-revving in direct top at this speed, or pulling just 4,650rpm in overdrive top. In the case of the V8, the 4.7-litre engine would have been turning over at 6,250rpm (and maximum power was developed at 6,000rpm).

Four-cylinder cars: There are few surprises in the comparison between models, but they confirm the fact that cars built after the establishment of the Lilley regime were no less lively, even though they were rather more heavy. For sure, the MGB engine/gearbox assembly must have been considerably heavier than the Ford powertrain, as a comparison between the Grantura Mark IVS and the Vixen S2 shows, while it is also quite clear that the adoption of the M-Series frame and body was responsible for another considerable increase in weight.

Six-cylinder and V8 cars: As may be expected, there was very little difference in the performance of any of the 3-litre (British) Ford V6-engined cars, and the latest 2.8 (German) Ford-engined Tasmin is not only faster than any of the normally aspirated 3-litre cars through the gears, but has better aerodynamics and a higher top speed.

Nothing more than a line of exclamation marks is needed to describe the performance of the Turbo-engined Convertible which *Autocar* tested in 1979 (no wonder their test sub-heading was 'Machismo lives...'). A figure-by-figure comparison with *Motor*'s Tuscan V8 is very interesting, for the 230bhp (DIN) Turbo was facing the 271bhp (gross) Tuscan V8. The lesson is that the bigger 4.7-litre engine of the Tuscan V8 had more torque, but that the 3-litre Turbo engine had rather more high-speed urge. Since both rushed up to 100mph in less that 15 seconds (a V12 Jaguar E-type tested by *Autocar*, incidentally, took 15.4 seconds) and both recorded between 16 and 18mpg fuel consumption, you 'paid yer money and took yer choice'. (For status, refinement and better engineering, mine would be the Turbo.)

'Federalized' US-market cars: The first point to note is that the car tested by *Road & Track* in January 1973 was placarded as a 2500M, but actually had the old-type bodyshell and therefore counts as one of the last of the 2500s. The second is that the maximum speed of the 1970-type Tuscan V8 (which had the short-run, widened bodyshell which was effectively the forerunner of the M-Series style) was restricted by the gearing and the valve–float limit (5,200rpm) of the non-standard 4,949cc Ford engine fitted to the test car. Incidentally, it is hard to believe that the V8 was producing its 220bhp (gross) for the performance figures are only slightly better than those quoted eight years later for the 135bhp (DIN) 3-litre Convertible.

	Grantura IIA 1588cc 4-cyl MG	Grantura IIA 1216cc 4-cyl C-C	Grantura 1800S 1798cc 4-cyl MG	Vixen S2 1599cc 4-cyl Ford	1600M 1599cc 4-cyl Ford	Tuscan V6 2994cc 6-cyl Ford	3000M 2994cc 6-cyl Ford
Mean maximum speed (mph)	98	101	108	109	105	125*	121
Acceleration (sec)							
0–30mph	4.4	4.2	4.4	3.6	3.2	2.6	2.5
0–40mph	5.9	6.0	6.3	5.1	4.9	3.8	4.0
0–50mph	9.3	8.2	8.4	7.4	7.2	6.0	5.8
0–60mph	12.0	10.8	10.9	10.5	10.4	8.3	7.7
0–70mph	17.2	15.2	14.3	14.0	14.3	11.1	10.6
0–80mph	23.7	20.7	18.9	18.8	19.6	14.3	14.0
0–90mph	–	27.8	25.5	26.4	27.8	18.7	18.8
0–100mph	–	–	36.8	41.7	40.7	24.8	25.6
0–110mph	–	–	–	–	–	33.5	37.0
0–120mph	–	–	–	–	–	–	–
0–130mph	–	–	–	–	–	–	–
Standing ¼-mile(sec)	19.0	18.3	18.0	17.2	17.6	16.2	16.0
Direct top gear (sec)							
10–30mph	9.8	–	–	–	–	–	–
20–40mph	8.9	–	8.5	9.3	10.4	6.7	6.4
30–50mph	9.0	10.9	8.9	9.1	10.4	6.6	6.6
40–60mph	9.8	11.4	9.6	9.3	11.8	6.3	6.4
50–70mph	11.8	13.4	10.2	8.7	9.5	6.1	6.1
60–80mph	14.6	13.7	10.9	10.2	13.3	6.6	7.4
70–90mph	–	16.1	13.1	12.4	15.7	7.9	8.7
80–100mph	–	–	19.4	22.5	22.9	9.6	10.9
90–110mph	–	–	–	–	–	15.4	16.3
100–120mph	–	–	–	–	–	–	–
110–130mph	–	–	–	–	–	–	–
Overall fuel consumption (mpg)	23.7	25.8	23.1	26.5	25.2	22.6	21.4
Typical fuel consumption (mpg)	33	30	29	28	28	27	24
Kerb weight (lb)	1680	1540	1904	1624	1974	1988	2240
Original test published	1961**	1961	1966**	1969	1976	1969**	1974

** Tested by *Motor*; all other British tests by *Autocar*
* Estimated speed in overdrive top gear

	Convertible Turbo	Tuscan	2500	2500M	Convertible	Tuscan
	2994cc 6-cyl Ford	4727cc V8 Ford USA	2498cc 6-cyl Triumph	2948cc 6-cyl Triumph	2994cc 6-cyl Ford	4949cc V8 (220bhp) Ford USA
Mean maximum speed (mph)	139	155***	111	109	125***	119****
Acceleration (sec)						
0–30mph	2.4	2.5	3.4	3.0	2.6	2.8
0–40mph	3.2	3.5	5.1	4.5	3.9	3.7
0–50mph	4.5	4.5	7.5	6.6	5.7	4.8
0–60mph	5.8	5.7	10.6	9.3	7.7	7.2
0–70mph	7.2	7.5	14.2	12.8	10.7	10.2
0–80mph	9.4	9.0	18.5	17.8	14.1	13.1
0–90mph	11.5	10.8	–	25.3	18.8	–
0–100mph	14.3	13.8	31.0	–	–	25.0
0–110mph	18.2	16.8	–	–	–	–
0–120mph	23.1	20.3	–	–	–	–
0–130mph	31.3	–	–	–	–	–
Standing ¼ -mile (sec)	14.5	14.1	17.7	17.3	16.6	15.4
Direct top gear (sec)						
10–30mph	–	–	–	–	–	–
20–40mph	8.2	5.0	–	–	–	–
30–40mph	7.1	4.8	–	–	–	–
40–60mph	6.8	4.9	–	–	–	–
50–70mph	6.4	4.9	–	–	–	–
60–80mph	5.7	4.9	–	–	–	–
70–90mph	5.0	5.0	–	–	–	–
80–100mph	5.3	5.6	–	–	–	–
90–110mph	6.3	6.3	–	–	–	–
100–120mph	8.3	6.9	–	–	–	–
110–130mph	13.2	–	–	–	–	–
Overall fuel consumption (mpg)	17.7	16.1	–	–	–	–
Typical fuel consumption (mpg)	19	17	32	29	25	20
Kerb weight (lb)	2436	2274	2150	2275	2335	2240
Original test published	1979	1967**	1973	1977	1978	1970

*** Estimated speed   **** Limited by 5,200rpm and axle gearing